Random Processes

STATISTICS: Textbooks and Monographs

, A Series Edited by

D. B. Owen, Coordinating Editor
Department of Statistics
Southern Methodist University
Dallas, Texas

R. G. Cornell, Associate Editor
for Biostatistics
University of Michigan

W. J. Kennedy, Associate Editor
for Statistical Computing
Iowa State University

A. M. Kshirsagar, Associate Editor
for Multivariate Analysis and
Experimental Design
University of Michigan

E. G. Schilling, Associate Editor
for Statistical Quality Control
Rochester Institute of Technology

Vol. 1: The Generalized Jackknife Statistic, *H. L. Gray and W. R. Schucany*
Vol. 2: Multivariate Analysis, *Anant M. Kshirsagar*
Vol. 3: Statistics and Society, *Walter T. Federer*
Vol. 4: Multivariate Analysis: A Selected and Abstracted Bibliography, 1957-1972, *Kocherlakota Subrahmaniam and Kathleen Subrahmaniam* (out of print)
Vol. 5: Design of Experiments: A Realistic Approach, *Virgil L. Anderson and Robert A. McLean*
Vol. 6: Statistical and Mathematical Aspects of Pollution Problems, *John W. Pratt*
Vol. 7: Introduction to Probability and Statistics (in two parts), Part I: Probability; Part II: Statistics, *Narayan C. Giri*
Vol. 8: Statistical Theory of the Analysis of Experimental Designs, *J. Ogawa*
Vol. 9: Statistical Techniques in Simulation (in two parts), *Jack P. C. Kleijnen*
Vol. 10: Data Quality Control and Editing, *Joseph I. Naus* (out of print)
Vol. 11: Cost of Living Index Numbers: Practice, Precision, and Theory, *Kali S. Banerjee*
Vol. 12: Weighing Designs: For Chemistry, Medicine, Economics, Operations Research, Statistics, *Kali S. Banerjee*
Vol. 13: The Search for Oil: Some Statistical Methods and Techniques, *edited by D. B. Owen*
Vol. 14: Sample Size Choice: Charts for Experiments with Linear Models, *Robert E. Odeh and Martin Fox*
Vol. 15: Statistical Methods for Engineers and Scientists, *Robert M. Bethea, Benjamin S. Duran, and Thomas L. Boullion*
Vol. 16: Statistical Quality Control Methods, *Irving W. Burr*
Vol. 17: On the History of Statistics and Probability, *edited by D. B. Owen*
Vol. 18: Econometrics, *Peter Schmidt*
Vol. 19: Sufficient Statistics: Selected Contributions, *Vasant S. Huzurbazar (edited by Anant M. Kshirsagar)*
Vol. 20: Handbook of Statistical Distributions, *Jagdish K. Patel, C. H. Kapadia, and D. B. Owen*
Vol. 21: Case Studies in Sample Design, *A. C. Rosander*
Vol. 22: Pocket Book of Statistical Tables, *compiled by R. E. Odeh, D. B. Owen, Z. W. Birnbaum, and L. Fisher*
Vol. 23: The Information in Contingency Tables, *D. V. Gokhale and Solomon Kullback*

Vol. 24: Statistical Analysis of Reliability and Life-Testing Models: Theory and Methods, *Lee J. Bain*

Vol. 25: Elementary Statistical Quality Control, *Irving W. Burr*

Vol. 26: An Introduction to Probability and Statistics Using BASIC, *Richard A. Groeneveld*

Vol. 27: Basic Applied Statistics, *B. L. Raktoe and J. J. Hubert*

Vol. 28: A Primer in Probability, *Kathleen Subrahmaniam*

Vol. 29: Random Processes: A First Look, *R. Syski*

Vol. 30: Regression Methods: A Tool for Data Analysis, *Rudolf J. Freund and Paul D. Minton*

Vol. 31: Randomization Tests, *Eugene S. Edgington*

Vol. 32: Tables for Normal Tolerance Limits, Sampling Plans, and Screening, *Robert E. Odeh and D. B. Owen*

Vol. 33: Statistical Computing, *William J. Kennedy, Jr. and James E. Gentle*

Vol. 34: Regression Analysis and Its Application: A Data-Oriented Approach, *Richard F. Gunst and Robert L. Mason*

Vol. 35: Scientific Strategies to Save Your Life, *I. D. J. Bross*

Vol. 36: Statistics in the Pharmaceutical Industry, *edited by C. Ralph Buncher and Jia-Yeong Tsay*

Vol. 37: Sampling from a Finite Population, *J. Hajek*

Vol. 38: Statistical Modeling Techniques, *S. S. Shapiro*

Vol. 39: Statistical Theory and Inference in Research, *T. A. Bancroft and C.-P. Han*

Vol. 40: Handbook of the Normal Distribution, *Jagdish K. Patel and Campbell B. Read*

Vol. 41: Recent Advances in Regression Methods, *Hrishikesh D. Vinod and Aman Ullah*

Vol. 42: Acceptance Sampling in Quality Control, *Edward G. Schilling*

Vol. 43: The Randomized Clinical Trial and Therapeutic Decisions, *edited by Niels Tygstrup, John M. Lachin, and Erik Juhl*

Vol. 44: Regression Analysis of Survival Data in Cancer Chemotherapy, *Walter H. Carter, Jr., Galen L. Wampler, and Donald M. Stablein*

Vol. 45: A Course in Linear Models, *Anant M. Kshirsagar*

Vol. 46: Clinical Trials: Issues and Approaches, *edited by Stanley H. Shapiro and Thomas H. Louis*

Vol. 47: Statistical Analysis of DNA Sequence Data, *edited by B. S. Weir*

Vol. 48: Nonlinear Regression Modeling: A Unified Practical Approach, *David A. Ratkowsky*

Vol. 49: Attribute Sampling Plans, Tables of Tests and Confidence Limits for Proportions, *Robert E. Odeh and D. B. Owen*

Vol. 50: Experimental Design, Statistical Models, and Genetic Statistics, *edited by Klaus Hinkelmann*

Vol. 51: Statistical Methods for Cancer Studies, *edited by Richard G. Cornell*

Vol. 52: Practical Statistical Sampling for Auditors, *Arthur J. Wilburn*

Vol. 53: Statistical Signal Processing, *edited by Edward J. Wegman and James G. Smith*

Vol. 54: Self-Organizing Methods in Modeling: GMDH Type Algorithms, *edited by Stanley J. Farlow*

Vol. 55: Applied Factorial and Fractional Designs, *Robert A. McLean and Virgil L. Anderson*

Vol. 56: Design of Experiments: Ranking and Selection, *edited by Thomas J. Santner and Ajit C. Tamhane*

Vol. 57: Statistical Methods for Engineers and Scientists. Second Edition, Revised and Expanded, *Robert M. Bethea, Benjamin S. Duran, and Thomas L. Boullion*

Vol. 58: Ensemble Modeling: Inference from Small-Scale Properties to Large-Scale Systems, *Alan E. Gelfand and Crayton C. Walker*

Vol. 59: Computer Modeling for Business and Industry, *Bruce L. Bowerman and Richard T. O'Connell*

Vol. 60: Bayesian Analysis of Linear Models, *Lyle D. Broemeling*

Vol. 61: Methodological Issues for Health Care Surveys, *Brenda Cox and Steven Cohen*

Vol. 62: Applied Regression Analysis and Experimental Design, *Richard J. Brook and Gregory C. Arnold*

Vol. 63: Statpal: A Statistical Package for Microcomputers – PC-DOS Version for the IBM PC and Compatibles, *Bruce J. Chalmer and David G. Whitmore*

Vol. 64: Statpal: A Statistical Package for Microcomputers – Apple Version for the II, II+, and IIe, *David G. Whitmore and Bruce J. Chalmer*

Vol. 65: Nonparametric Statistical Inference, Second Edition, Revised and Expanded, *Jean Dickinson Gibbons*

Vol. 66: Design and Analysis of Experiments, *Roger G. Petersen*

Vol. 67: Statistical Methods for Pharmaceutical Research Planning, *Sten W. Bergman and John C. Gittins*

Vol. 68: Goodness-of-Fit Techniques, *edited by Ralph B. D'Agostino and Michael A. Stephens*

Vol. 69: Statistical Methods in Discrimination Litigation, *edited by D. H. Kaye and Mikel Aickin*

Vol. 70: Truncated and Censored Samples from Normal Populations, *Helmut Schneider*

Vol. 71: Robust Inference, *M. L. Tiku, W. Y. Tan, and N. Balakrishnan*

Vol. 72: Statistical Image Processing and Graphics, *edited by Edward J. Wegman and Douglas J. DePriest*

Vol. 73: Assignment Methods in Combinatorial Data Analysis, *Lawrence J. Hubert*

Vol. 74: Econometrics and Structural Change, *Lyle D. Broemeling and Hiroki Tsurumi*

Vol. 75: Multivariate Interpretation of Clinical Laboratory Data, *Adelin Albert and Eugene K. Harris*

Vol. 76: Statistical Tools for Simulation Practitioners, *Jack P. C. Kleijnen*

Vol. 77: Randomization Tests, Second Edition, *Eugene S. Edgington*

Vol. 78: A Folio of Distributions: A Collection of Theoretical Quantile-Quantile Plots, *Edward B. Fowlkes*

Vol. 79: Applied Categorical Data Analysis, *Daniel H. Freeman, Jr.*

Vol. 80: Seemingly Unrelated Regression Equations Models : Estimation and Inference, *Virendra K. Srivastava and David E. A. Giles*

Vol. 81: Response Surfaces: Designs and Analyses, *Andre I. Khuri and John A. Cornell*

Vol. 82: Nonlinear Parameter Estimation: An Integrated System in BASIC, *John C. Nash and Mary Walker-Smith*

Vol. 83: Cancer Modeling, *edited by James R. Thompson and Barry W. Brown*

Vol. 84: Mixture Models: Inference and Applications to Clustering, *Geoffrey J. McLachlan and Kaye E. Basford*

Vol. 85: Randomized Response: Theory and Techniques, *Arijit Chaudhuri and Rahul Mukerjee*

Vol. 86: Biopharmaceutical Statistics for Drug Development, *edited by Karl E. Peace*

Vol. 87: Parts per Million Values for Estimating Quality Levels, *Robert E. Odeh and D. B. Owen*

Vol. 88: Lognormal Distributions: Theory and Applications, *edited by Edwin L. Crow and Kunio Shimizu*

Vol. 89: Properties of Estimators for the Gamma Distribution, *K. O. Bowman and L. R. Shenton*

Vol. 90: Spline Smoothing and Nonparametric Regression, *Randall L. Eubank*

Vol. 91: Linear Least Squares Computations, *R. W. Farebrother*

Vol. 92: Exploring Statistics, *Damaraju Raghavarao*

Vol. 93: Applied Time Series Analysis for Business and Economic Forecasting, *Sufi M. Nazem*

Vol. 94: Bayesian Analysis of Time Series and Dynamic Models, *edited by James C. Spall*

Vol. 95: The Inverse Gaussian Distribution: Theory, Methodology, and Applications, *Raj S. Chhikara and J. Leroy Folks*

Vol. 96: Parameter Estimation in Reliability and Life Span Models, *A. Clifford Cohen and Betty Jones Whitten*

Vol. 97: Pooled Cross-Sectional and Time Series Data Analysis, *Terry E. Dielman*

Vol. 98: Random Processes: A First Look, Second Edition, Revised and Expanded, *R. Syski*

Vol. 99: Generalized Poisson Distributions: Properties and Applications, *P.C. Consul*

ADDITIONAL VOLUMES IN PREPARATION

Random Processes

A FIRST LOOK

Second Edition
Revised and Expanded

R. Syski

Department of Mathematics
University of Maryland
College Park, Maryland

CRC Press
Taylor & Francis Group
Boca Raton London New York

CRC Press is an imprint of the
Taylor & Francis Group, an **informa** business

First published 1989 by Marcel Dekker, Inc.

Published 2019 by CRC Press
Taylor & Francis Group
6000 Broken Sound Parkway NW, Suite 300
Boca Raton, FL 33487-2742

© 1989 by Taylor & Francis Group, LLC
CRC Press is an imprint of Taylor & Francis Group, an Informa business

First issued in paperback 2019

No claim to original U.S. Government works

ISBN 13: 978-0-367-45119-6 (pbk)
ISBN 13: 978-0-8247-8028-9 (hbk)

Visit the Taylor & Francis Web site at
http://www.taylorandfrancis.com

and the CRC Press Web site at
http://www.crcpress.com

To inseparable brothers

Buc and Mieszko,

who cherish random events

Foreword

Many disciplines have their beginnings in the arcane struggles of a small group of researchers. When the ideas generated are seen to relate to human problems, they spread to a larger, less specialized group. Then new questions develop, stimulating both the growth of the discipline and its further spread. Through such success and involvement, mathematics has penetrated and shaped the physical and engineering sciences, growing itself thereby. Even the softer sciences of biology, medicine, psychology, economics and management, traditionally qualitative, have participated in this process, and today respond increasingly to the demands of mathematics and its benefits.

In all these sciences, chance plays a key role, and the special branch of mathematics devoted to its study, probability theory, intrudes itself. One area of probability theory, statistical inference, deals largely with the implications of randomness in fixed data for estimation and prediction. A second and somewhat newer area, stochastic processes, deals with the dynamics of change in the presence of

randomness. It studies entities fluctuating randomly in
time such as temperature levels, inventory levels, air indices, noise levels, delays in a time-shared computer. Such
processes have been studied in all the sciences mentioned
above and are of growing importance.

Probability theory is one of the more difficult
branches of mathematics in that its ideas are subtle and
sometimes confusing. When this natural difficulty is aggra-
vated by a formal, abstract exhaustive exposition of the
subject, it is placed beyond the reach of many of the people
who need its tools. There has been a great need to present
the basic ideas in a simple, lively manner, with the formal
trappings of mathematics suppressed. As Professor Syski
demonstrates, probability theory can be entertaining and
does not have to be hidden in intellectual thickets.

The author is well qualified technically to write
the kind of book needed. He is a highly competent mathematician and probabilist and has published extensively in the
research literature. He has contributed to probabilistic
potential theory, a very abstract and pure branch of probability theory. He has also contributed to the applied
literature. His book on congestion theory was a pioneering
work in this field, of value to both practitioners and
theoreticians alike.

There are, however, many good probabilists, pure,
applied and both. Two special human qualities are required
to write such a book, vitality and humor. That Professor

Syski has these qualities has long been known to his friends,
and his astonishing enthusiasm for life, people and probability theory are very evident in this book.

> J. Keilson
> University of Rochester
> Rochester, New York

Preface to the Second Edition

It has been a great pleasure for the author to see that this book was well received. Indeed, initial reviews by experts in the field were favorable (with, of course, the usual amount of constructive criticism). Several letters from individual persons who read the book and enjoyed it were especially gratifying. It seems that the novelty of this "first look" was generally appreciated.

This new edition contains several modifications; however, the spirit and the character of the book remain unchanged. The misprints have been corrected and some confusing statements (pointed out by readers) have been clarified by rewriting and by adding new material. The following are more substantial modifications.

In Chapter 3, Section 17 on renewals has been amplified. Extensive additions have been made in Chapter 4. Section 25 on the birth-and-death process has been enlarged by the addition of Subsection 25.4 dealing with the first-entrance concept, and the completely new Section 26 on

branching process has been added. Moreover, in all chapters
the number of problems (taken from examination papers) was
considerably increased.

Finally, a new Chapter 5 on statistical estimation
has been added. This perhaps requires a word of justification. Apart from the natural connection between Probability
and Statistics, the statistical analysis of random processes
is of recent origin. It may therefore be desirable to discuss these new procedures within the limits imposed by the
character of this book. The discussion is thus restricted
to material presented earlier in the book and actually
treats only the maximum likelihood method, as most suitable
from the analytical point of view. Apart from the introductory statistical Sections 27 and 28, the chapter contains
material that is not seen in elementary texts. Thus, Sections 29 and 30 formally are in the field of statistics but
actually continue the presentation of stochastic processes.

The last section presents some concluding after-
thoughts. It offers general comments on the art of probabilistic modeling—its advantages and pitfalls—and invites
readers to further study of the fascinating field of random
processes. Several recent references have been added to the
Suggestions for Further Reading (Appendix C).

The book is the outgrowth of a course on Stochastic
Models (Stat. 250) offered by the Statistics Branch, Department of Mathematics, University of Maryland, as a first
introduction to applications of probability (with aims mentioned in the Preface).

The author wishes to express his gratitude to his colleagues in the Statistics Branch, associated with the development of the Stat. 250 course, for their valuable comments. The book also owes much to the reaction of students in the Stat. 250 classes.

The author is grateful to Julian Keilson for his interest and his kind words and shares with him his views on the role of applied probability.

My warm thanks go to my son Marek, who prepared the illustrations and drawings, and I am indebted to Pat Berg for her superb typing of the manuscript for the first edition.

It is a pleasure to thank Marcel Dekker, Inc., for its interest in the publication of this book.

R. Syski

Preface to the First Edition

It is common knowledge that observable phenomena in the real life world -- from playing of games to the rise and fall of empires -- are governed in their development to a large extent by chance. The study of the underlying chance mechanisms lies in the domain of Probability Theory, and is called the theory of random (or stochastic) processes. It is to the credit of this theory that it can describe rather accurately behavior of simple or complex systems, composed of living organisms (human beings, animals, bacteria) or material objects (machines, cars, equipment, etc.), subjected to chance fluctuations. These fluctuations have various origins both internal and external to the structure under consideration, and are also related to the complexity of such a system.

Substantial progress has been made in recent years in such areas as reliability (dealing with survival questions of systems and their components), queueing (dealing with familiar phenomena associated with waiting), traffic (telephone, road, air, sea), inventories and storage (supply and demand fluctuations), medical aspects (spread of epidemics, survival and treatment effectiveness), psychology (learning models), economics and social sciences (human behavior).

The theory of stochastic processes owes its success to solid mathematical foundations provided by modern probability which itself made "a great leap" from gambling to respectability. Nowadays, the theory of random processes is a well-established branch of mathematics with an extensively developed theoretical side (a paradise for abstract theoreticians) and with impressive applications (for the practically minded researchers). Numerous volumes have been written in this field, ranging from elementary texts to sophisticated monographs, and a beginner must invest a great effort of time, patience and determination to reach even the "first degree of initiation."

There are, however, numerous people who for various reasons either cannot afford or do not wish to undergo such demanding effort, but who would like to know for their own enlightment what the theory of stochastic processes is, and who have (perhaps moderate) mathematical background sufficient to appreciate what is shown to them.

To these people this book is addressed. Its prerequisite is an interest in the subject and a working knowledge of elementary calculus. The book is indeed "the first look" at random processes. It is not a systematic account of the theory (such accounts can be found in existing textbooks), but takes a light-hearted yet correct and moderately precise approach to selected highlights of the theory. The

book does not offer a shortcut to the mastery of the subject
(no such shortcuts are possible!), but it presents only
convincing arguments (without rigors of formal proofs) in
justification of the announced results. Indeed, the book
aims to develop appreciation of the ingenuity involved in
the mathematical treatment of random phenomena, and of the
power of the mathematical methods employed in the solution
of applied problems. It is hoped that this book will stimu-
late readers' interest in further study of stochastic processes, by perhaps showing the applicability of methods and
results to the readers' own field of interest. (Even if it
fails in this respect, the casual reader who only glanced
through its pages would benefit by gaining a new vocabulary.)

****

A large group of readers to whom this book is addressed are students not majoring in mathematics, but
interested in applications of probability to their disciplines.

It may be desirable for non-math majors who have
just completed a semester of calculus to see applications to
the "real life" problems in social sciences, operations
research and statistics. Such applications may have more
appeal to some students than the routine examples from
physics and engineering (usually found in calculus textbooks). This book has been designed to perform such a task
with a two-fold aim:

i) to expose students to basic concepts of stochastic processes in an informal way through an intuitive approach
with the help of calculus,

ii) to provide novel illustrative examples of applications
of calculus to practical problems.

The primary object being applications, the book concentrated on the discussion of several selected problems.
Selection has been made on individual merits, with problems
leading to interesting mathematical techniques having first
priority. However, topics extensively discussed in elementary literature have been usually omitted; instead many
problems from advanced texts have been deliberately included.

Each problem is stated first in plain language.
Then its probabilistic model is developed, equations written
down and solved (or only the solution announced), and finally
the resulting formulae for probabilities or mean values are
discussed and interpreted in practical terms. This essentially is reduced to evaluation of integrals (integration by
parts!), differentiation (chain rule!), discussion of properties of functions (slope, maxima and minima, inflection
points, limits, etc.), and solution of linear differential
equations with constant coefficients. From a probabilistic
point of view, readers get acquainted with such concepts as
the distribution function, density, random variables, expectation, moments, etc.

No formal proofs are given, but only justification
of very loosely stated assertions. A justification is selected if it presents an interesting example of calculus tech-

niques; otherwise, the result is shamelessly verified on one
or two examples and declared to be valid, especially if the
final result admits convincing practical interpretation.
Numerical calculations are strongly encouraged.

Although at first glance the problems may appear
unrelated to each other, their common bond is supplied by
their probabilistic background. Indeed, the prevailing
theme is the distribution of various life times in renewal
theory, reliability and Markov chains.

The material is arranged into four chapters, according to thematic affinity. Chapter 1 (Easy Life and Good
Times) is devoted to properties of life times, and tacitly
introduces basic probabilistic concepts. Chapter 2 (Be
Discreet with Discrete) centers around Bernoulli trials and
Poisson distribution, and their common applications. Chapter 3 (To Renew or Not to Renew) describes renewal theory
and its ramifications. This chapter is more advanced than
the preceding two, and requires more attention from the
reader. Chapter 4 (Markovian Dance) treats essentially
examples of birth and death processes with applications to
queueing and learning theory. Problems are treated separately and in order to stress applications at this level,
full power of Markovian theory is regretfully not utilized.
Each chapter begins with a short introductory comment. The
appendixes contain a list of formulae and tables. On the
whole, the book contains only classical material, and therefore the individual references are not given, but the interested reader should consult the references listed in Appendix
C (Suggestions for Further Reading).

To the student

This book aims to put some life in calculus techniques you learned earlier, and which may appear to you rather dry.

It is also intended to show you that many real life situations can be analyzed with the help of probabilistic arguments using simple calculus.

Do not be horrified by strange looking formulae on the following pages. Many of them are already familiar to you, although they may appear here in disguise. Others will become your friends soon, as you will study waiting times, queueing problems, learning, etc.

To the instructor

This is NOT a text in probability theory, nor is it a routine introduction to stochastic processes. This is simply a collection of several interesting problems which may appeal to imagination. Basic probabilistic concepts are introduced as they are needed, and in an unorthodox manner. Probability has been simply defined as the integral of density (area under a curve), or as a sum. A few new names have been invented to convey an idea (like "jumping rabbit" for a stopping time, "life times" for positive random variables, "cost function" for a function of a random variable).

If you have reservations against using such techniques like convolution integrals, double integrals, simple differential equations, etc., at this level, rest assured that they can be easily introduced in a very convincing manner. All that is needed is to tie them with the renewal

process, combination of life times, birth and death processes, reliability, the Poisson process, and so on, in a very
natural and simple way. Perhaps, using this simple interpretation of calculus concepts, the teaching of calculus
may be more attractive.

It is surprising to see how such concepts can be
appreciated by students at this level, and operated in a
satisfactory manner, without heavy formal preparation.

R. Syski

Contents

Foreword   v
Preface to the Second Edition   ix
Preface to the First Edition   xiii

1. Easy Life and Good Times   1

    1: Probability   2
    2: Life Times   9
    3: Prolongation of Life Times   22
    4: Bus Problem   29
    5: Combinations of Life Times   34
    6: Extreme Life Times   45
    7: Great Expectations   51
    8: Double Scotch   67
    9: How Normal Is Normal?   75
   10: When in Doubt, Approximate!   85
        Problems   91

2. Be Discreet with Discrete   131

   11: Bernoulli Trials   132
   12: Applications of Binomial   141
   13: Geometric Waiting Time   146
   14: Poisson Distribution   152
   15: Accidents Just Happen   157
        Problems   165

3. To Renew or Not to Renew   187

   16: Renewals   188
   17: Renewal Equation   195
   18: Jumping Rabbit   206
        Problems   214

4. Markovian Dance   231

   19: Poisson Input   233
   20: It Is Easy to Learn   241
   21: On-Off Transitions   248
   22: Queueing   256

23: Waiting Time   266
24: Birth Right   274
25: Birth and Death   283
26: Branching Process   303
     Problems   323

5.   Inference from Interference   341

27: Esteemed Estimators   344
28: As You Like It   349
29: Estimating Poisson Input   356
30: Estimating Birth and Death Process   361
31: Modeling   376
     Problems   387

Appendix A Formulae   393
Appendix B Tables   399
Appendix C Suggestions for Further Reading   405
Index   409

Random Processes

1
Easy Life and Good Times

In this chapter we shall discuss random fluctuations of life times associated with people, animals and machines. As we adopt a probabilistic point of view, it is appropriate to begin with a precise formulation of our intuitive notion of probability, and establish rules for calculations. Next, we shall define what we mean by life time, and state how it can be described in a convincing way. Then, we shall engage in the study of numerous examples of diverse nature, involving one life time, combinations of life times, and highly interesting associated features like cost functions, averages, distribution functions, densities and other normal and abnormal situations. Proceeding step-by-step, we shall develop our probabilistic tools as the need arises. It is the rewarding applications which will gratify our efforts in acquiring these tools. So let us hope for easy life and good times!

Perhaps we should stress again that in the probabilistic description of the real life situations, we shall be rather concerned with a mathematical model of the situation. The

real situation is too complex, and our tools rather crude,
so we must strike a balance between complexity of the situation and the complexity of our analysis. It should always
be remembered that our models are as good as the conclusions
which we can draw from them. If these conclusions are at
variance with observed facts, our model and our analysis are
of dubious validity.

Another fact should be stressed, too. We shall deal
with mathematical models and we shall need calculus. Para-
phrasing the inscription of the Platonian Academy, we can
say that there is no chance of studying probability without
calculus. So let us not hear complaints that what we are
doing is "just calculus and nothing but calculus!"

Section 1: Probability

Every person has some intuitive interpretation of the
meaning of the concept of probability (pr.). We may ask
such questions as, "what is the pr. that it will rain tomorrow," "what is the pr. that my favorite team will win a
particular game," "what is the pr. that I shall be killed
in a car accident," "what is the pr. that the waiting time·
for a bus is less than 10 minutes," and so on.

When speaking about pr., the usual picture formed in
one's mind is that of dice throwing, card playing, or any
other games of chance. Although it is true that Probability
Theory originated from such considerations, today this theory is a well developed, self-contained mathematical disci-

pline with many applications to the physical sciences,
engineering and the social sciences.

We shall be concerned here with a wide range of applications, ranging from the study of complicated systems of
machines and people to the study of human behavior. We
shall deal with topics which nowadays are generally called
Operations Research. It is a far cry from dice and cards;
as a matter of fact we shall ignore cards and dice almost
entirely. Problems which we are going to discuss are important real life problems, studied extensively in the literature on probability.

Unfortunately, we must restrict ourselves to the
simplest situations which can be handled with the limited
mathematical tools at our disposal. We need rudimentary
calculus -- this is a fact of life. Speaking about Probability without calculus is like taking a correspondence course
in driving a car.

Intuitively, pr. is regarded as the opposition of cer-
tainty. We do not say that an event will take place (or
took place), but we say only that such an event will (or
did) probably occur. The pr. expresses our lack of information, our uncertainty, our degree of belief. Indeed,
with all the evidence being the same, answers to the questions stated above may be entirely different, according to
who is making the estimate of pr. Just think about the pr.
of some controversial issue!

It will be necessary for us to eliminate from the start
any reference to individuals who make estimates of pr. Thus,

the pr. of an event will be the same, irrespective of who
calculated it.   This is done for purely technical reasons;
we shall, however, evaluate pr.'s of some events pertaining
to individuals, but we shall reformulate them appropriately.
In other words, we shall not discuss the subjective pr.,
but shall consider only the objective pr.

In comparing pr.'s of events, we may limit ourselves
to statements expressing the relative magnitude.   Thus, we
may wish to know only that the pr. of one event is larger
than that of some other event.   However, it is much more
natural and convenient, to express the pr. of an event by a
number.   One can then operate with such numbers.   For example, suppose that the pr. of waiting for a bus less than 10
minutes is   1/2,   whereas for waiting less than one hour is
almost 1.   How have these numbers been obtained?   What do
they mean?   How can one use them?   That's what this book is
about.

The idea of attaching a number to a set or to an object
is familiar to everybody.   We may talk about the length of a
segment of a line, the area of a triangle, the volume of a
sphere, or the mass of a physical body, its temperature, its
price, and so on.   We may speak about the number of people
in the waiting room, the number of bacteria in a culture,
the time needed to memorize a particular poem, the number
of miles per gallon, the speed, the acceleration, the temperature, the pressure; the gain and the loss, the life and
death, pleasure and pain, the beauty and the beast -- all
these facts can be expressed in numbers.

Probability is also expressed in terms of numbers attached to events. The way of doing this is very much analogous to that of length, area and volume -- keep this in mind! Like these quantities, the pr. is nonnegative. If you have two plane figures which are disjoint, their total area is equal to the sum of their respective areas; if a farmer has 500 acres and his neighbor has 600 acres, their total holdings are 1100 acres. The volume of a sphere is equal to, say 15 cubic inches; any part of the ball necessarily has volume smaller than 15 cubic inches. It may be convenient to rescale the ball by saying that the volume is 1; then any part has volume expressed in fractions, or in percentage. Similarly, with the pr. One says that the pr. of an event which includes all events, the so called universal event or plainly the total event, is 1. Then, the pr. of any event is expressed by a fraction or a percentage. In other words, the pr. of a sure event (i.e., the event we know is bound to occur) is 1; the pr. of any other event lies therefore between 0 and 1. The exclusive events, that is events which cannot occur simultaneously are treated like disjoint sets; their probabilities add. Thus, if the pr. of having 5 people in the waiting room is 1/2 and the pr. of having 10 people is 1/3, the pr. of having either 5 or 10 people is 1/2 + 1/3 = 5/6. If the waiting room can hold 15 people only, the pr. of having 15 or less is 1. Note that the pr. of this room holding 20 people is 0, but of holding less than 20 is also 1.

In order to achieve a generality of discussion that is independent of the nature of events under consideration, it is convenient to denote events by letters (usually capitals) such as A, B, C etc. With our abbreviation of probability already introduced, we shall write the pr. of an event E in the form:

$$pr(E).$$

Thus, instead of the lengthy statement that "the waiting time for something is less than one hour with pr. 1/4," we shall write

$$pr(E) \quad = \quad 1/4$$

with the understanding that E represents the event in question. In general, for any event A

$$0 \quad \leq \quad pr(A) \quad \leq \quad 1.$$

Also, if events A and B are mutually exclusive, then

$$pr(A \text{ or } B) \quad = \quad pr(A) + pr(B).$$

In particular, if A is an arbitrary event, then A^c will denote the complementary event; that is A^c means that A does not occur. For example, if A represents the event that there are exactly 6 machines in operation, the event

A^c means that the number of machines in operation is not
6. Since A and A^c are obviously mutually exclusive,
one has the following useful relation:

$$pr(A) + pr(A^c) = 1.$$

How does one compute such numbers as $pr(A)$? There are
certain rules and methods designed for this purpose and we
shall talk about them as we solve several practical problems.
Perhaps it may be of some interest to mention a rule which
used to be of some importance in olden days, but has later
been relegated to a remote place; yet it has some intuitive
appeal and you may find it being used from time to time.
Suppose that you wish to perform some simple experiment,
like tossing a coin or a die, which may result in several
outcomes. Suppose further that you are interested in a
particular outcome only, say the head on a coin, or the 6
on a die. You perform this experiment many times and count
the number of occurences of that particular outcome. Then,
the pr. of the event A -- that is, the occurence of the
outcome in question -- is a ratio of the number of occurences of that outcome to the total number of repetitions of
the experiment. For example, if you toss a coin 100 times
and the number of times a head appears is 57, then the
pr. of a head is taken as 57%.

Such a procedure is rather vague, but supplies a pic-
tureque intuitive background; it is known as the frequency

approach. It is very popular in applications, especially
in engineering. It fails miserably when the nature of
things prevents repetitions of the experiment. On the other
hand, under very special conditions (which we shall not
discuss here) one can show that when the number of repetitions of the experiment is sufficiently large, the frequency
approach will agree with the approach already mentioned
above. This is known as a law of large numbers.

For us, however, the pr. of an event is analogous to
length, area or volume; it is a number associated with an
event. In our discussions here, we shall investigate practical problems in which we shall be able to compute probabilities directly. Because of the highly involved structure of
real life situations, we shall restrict ourselves to investigations of rather simple situations. In other words, we
shall simplify our task by considering simplified versions
of the real life problems; that is, we shall study
probabilistic models. These models will be mathematically
simple, yet they will be sufficiently realistic.

We shall make some assumptions about the structure of
our model, and then derive several expressions for various
probabilities of interest. This will be the probability
part. As a result, we shall get a formula. To contrast it
with the real life situation, we shall use data collected
from observations, from experiments, etc., and shall draw
appropriate conclusions. This will be the statistics part.

Section 2: Life Times

The presence of a time factor is a distinct feature of many real life situations. When buying new equipment, say a car, a radio or a dress, one usually asks how long it will last. Sitting in the waiting room of a dentist, standing in a queue in front of a ticket office, waiting for a bus -- all these are familiar situations involving waiting time.

More generally, we shall call a time needed to perform some function, to observe something, a time for something to happen -- a life time. This may be the actual life time of equipment, a life time of a human being or of an animal; or a waiting time for the arrival of a plane, a duration of a telephone conversation, travel time between two cities, and so on.

In all these situations there is an element of chance involved. Indeed, the length of the life time fluctuates, depending on many factors. Although one may be interested in the exact duration of a life time, in most situations it is enough to have some approximate estimate of this time. You may be sceptical when told that a new gadget you just bought will last exactly 30 days; you will be more convinced when told that it will probably last about a month. Similarly, it is very likely that your new car will perform well for about a year, and then repairs may ruin you; or with bad luck somebody may hit your car the very moment you leave the dealer's premises. Thus the exact length of time

your new car will operate until the first major breakdown
may be hard, or even impossible, to determine. You would
rather prefer to have some approximate estimate, say "about
a year;" or, checking published reports, you may find that
"on the average" that model of a car has such and such frequency of repairs.

Thus, it would be very desirable to have some way of
expressing quantitatively these possible fluctuations of
life times. It is here where Probability Theory enters.
Indeed, we shall talk about the pr. that a life time lies
within specified limits. For example, what is the pr. that
your new car will perform well for more than a month, for
more than a year? What is the pr. that within the length of
time you own this car, there will be one, two, ten major
repairs?

In other examples, one may ask for the pr. that the
waiting time for the first occurence of some event will be
larger than a specified amount? What is the pr. that a
telephone conversation will terminate within ten minutes
(when you are waiting outside the booth)?

In order to provide answers to such questions, it is
necessary to examine more closely the notion of a life time.
It will be convenient to write simply X for the life time.
Thus, X is a variable quantity whose fluctuations depend on
chance. Hence, our life time X is an example of a <u>random</u>
<u>variable</u>, a notion essential for Probability Theory.

We observe that the life time X is a positive quantity; it may be zero, of course. However, we shall ignore

a possibility of negative life times, although random variables assuming positive and negative values are very common
in other applications.

It is clear that a life time is finite, but we may
hesitate to impose any definite bound. A man can live 100
years, but not 200; what about 101, 102, ... ? 150? However, for convenience we shall assume that the life span is
finite, say of length L which may be one minute, ten years
or whatever is the case. Furthermore, we shall assume that
the life time X starts at some instant of time, "the origin," which will be denoted by s; say, today, tomorrow at
10 p.m., two years ago, and so on. Thus, our particular
life time X begins at the moment s and continues until
the instant s + L (when it terminates), its total duration
being L.

If the life time X is exactly of length t, where t
is a number between s and s + L, say t seconds, minutes,
years, etc., we shall write $(X = t)$; here the symbol $(X =
t)$ denotes the event that the life time X equals exactly
t. This type of symbolic notation is very economical. For
example, the event that the life time X has duration between a and b, inclusively, can be simply written in our
shorthand notation as:

$$(a \leq X \leq b).$$

Here a and b are numbers such that a is not larger
than b, and both are included between s and s + L. For
example, suppose that X represents a waiting time, being

measured from the instant zero until one hour; thus s = 0
and L = 60 (minutes). The event that one waits from 5
to 15 minutes is then written as (5 ≤ X ≤ 15). When
there is no possibility of confusion, we may even drop X
from our notation and denote the event in question simply
by [a,b]. In other words, we expressed our event concerning the duration of the waiting time, by an interval equal
to that duration. This is a very convenient procedure,
and we shall employ it frequently.

We now must associate pr. with the events concerning
X. In view of what has been said about probability in the
previous section, one must associate numbers with intervals
representing events. We are at liberty in choosing these
numbers in any reasonable way. At this stage, it is necessary to distinguish between two important modes of counting
time. We may measure time in discrete units, say 1, 2, 3
etc. (minutes, years, centuries), or continuously (in agreement with a poetic expression that "time flows"). These two
methods require different mathematical treatment.

We shall now restrict our discussion to the continuous
case. We shall express pr.'s of events with the help of a
function f, called the <u>density</u> of life time X. This
function f is defined on the interval [s,s+L] and has
the following two properties:

i) $f(t) \geq 0$ for every t such that $s \leq t \leq s + L$.

ii) $\int_{s}^{s+L} f(t) \, dt = 1.$

Property (i) means that f is never negative; property (ii) indicates that the area under the curve f(t) is equal to 1. Otherwise the function f is quite arbitrary, and we shall soon see several common examples.

With density f at our disposal, we shall define the pr. of the event that the life time X takes values between a and b, inclusively, as the integral of the density f taken from a to b; in symbols:

$$pr(a \leq X \leq b) \;=\; \int_a^b f(t)\ dt, \qquad where \;\; s \leq a \leq b \leq s+L.$$

Thus, the pr. of the interval [a,b] is just the area under the curve f(t) from a to b. In other words, we associated a number -- equal to the area under the curve -- to an interval.

Observe that by property (ii), the total area is 1 and this corresponds to the fact that the life time X must take some value between s and s + L, inclusively:

$$pr(s \leq X \leq s+L) \;=\; 1.$$

It follows from property (i) and properties of integrals that for intervals [a,b] ⊂ [a',b'], one would have

$$pr[a,b] \;\leq\; pr[a',b'].$$

Similarly, if A and B are two disjoint intervals, then

$$pr(A \text{ or } B) = \int_A f(t)\ dt + \int_B f(t)\ dt \quad \text{the integration being}$$

taken over intervals A and B.

Of special interest are intervals of the form [s,t], where t varies from s to s + L, inclusively, and s is the initial point of the life time X. For this case we shall write:

$$pr(s \leq X \leq t) \quad = \quad \int_s^t f(x) \; dx, \qquad s \leq t \leq s + L.$$

Having fixed s and L, we may look at this integral as a function of t; denoted by F -- this function F is called the <u>distribution function</u> (d.f.) of a random variable (the life time) X. Thus,

$$F(t) \quad = \quad \int_s^t f(x) \; dx, \qquad s \leq t \leq s + L$$

and the number F(t) represents the pr. that the life time X is t or less; moreover, F(t) is equal to the area under the density curve taken from the origin s up to the point t.

Clearly,

$$0 \leq F(t) \leq 1, \quad \text{and} \quad F(s) = 0, \quad F(s+L) = 1.$$

Moreover, the d.f. F is always nondecreasing: $F(t) \leq F(t')$ whenever $t \leq t'$, because the interval [s,t] is contained in the interval [s,t'].

Observe also that

$$pr(a \leq X \leq b) \quad = \quad F(b) - F(a).$$

Of special interest is the event $(t < X \leq s+L)$, which is the complement of the event $(s \leq X \leq t)$. Indeed:

$$(s \leq X \leq t) \; \cup \; (t < X \leq s+L) \;\; = \;\; (s \leq X \leq s+L).$$

Hence, as we already noted in Section 1:

$$\text{pr}(t < X \leq s+L) \;\; = \;\; \int_{t}^{s+L} f(x) \, dx \;\; = \;\; 1 - F(t)$$

and this represents the area under the density curve from point t until the end. The expression $1 - F(t)$, denoted also by $F^{c}(t)$, is called the complementary d.f. in general, and the <u>survivor function</u>, with reference to the life time. Indeed, $F^{c}(t)$ gives the pr. that the life time has not failed up to time t. Clearly, $F^{c}(s) = 1$ and $F^{c}(s+L) = 0$; moreover, F^{c} is a nonincreasing function of t.

Observe that if the life time d.f. F is given, its density can be obtained by differentiation (for those t where f is continuous):

$$f(t) \;\; = \;\; dF(t)/dt$$

and similarly:

$$f(t) \;\; = \;\; -dF^{c}(t)/dt.$$

It should be stressed that $f(t)$ itself does not represent a pr.; indeed, the values of f may be larger than 1 for some t. However, the quantity $f(t) \, dt$ may be used to

represent approximately the pr. that X assumes values
between t and t + dt:

$$pr(t \leq X \leq t+dt) \quad \approx \quad f(t) \ dt;$$

the above expression is very useful in applications to practical problems.

Finally, observe that in the present case (when density
f is given) the pr. that the life time X assumes exactly
the value t is zero: indeed

$$pr(X = t) \quad = \quad \int_{t}^{t} f(x) \ dx \quad = \quad 0.$$

This is precisely the characterization of the continuous
random variables X we now discuss.

It is now time to stop for a few examples.

Example 1: Uniform distribution.

$$f(t) \quad = \quad \begin{cases} \dfrac{1}{L} & \text{for} \quad s \leq t \leq s+L \\ \\ 0 & \text{otherwise} \end{cases} \quad ;$$

$$F(t) \quad = \quad \int_{s}^{t} \frac{1}{L} \ dx \quad = \quad \frac{t-s}{L} \qquad \text{for} \quad s \leq t \leq s+L;$$

$$F^{c}(t) \quad = \quad 1 - \frac{t-s}{L} \ ;$$

$$pr(a \leq X \leq b) \quad = \quad \int_{a}^{b} \frac{1}{L} \ dx \quad = \quad \frac{b-a}{L} \ .$$

Example 2: Negative exponential distribution (n.e.d.).
(s = 0, L = ∞).

$$f(t) = \lambda e^{-\lambda t}, \quad t \geq 0, \quad \lambda > 0 \quad \text{constant;}$$

$$F(t) = \int_0^t \lambda e^{-\lambda x} \, dx = 1 - e^{-\lambda t}, \qquad t \geq 0;$$

$$F^c(t) = e^{-\lambda t}, \qquad t \geq 0;$$

$$pr(a \leq X \leq b) = \int_a^b \lambda e^{-\lambda x} \, dx = e^{-\lambda a} - e^{-\lambda b}, \qquad a \leq b.$$

Important simplification: It is a bit of a nuisance to
remember the values of the initial instant s and of the
duration L. In some cases we may take s = 0 (and we
shall frequently do so), and we shall also encounter life
times for which L is infinite. It is therefore mathematically more convenient to extend definitions of the d.f. F
and its density f to the whole nonnegative real line,
that is for all t from 0 to infinity. We shall simply
write:

$$F(t) = 0 \quad \text{for} \quad 0 \leq t \leq s$$

and

$$F(t) = 1 \quad \text{for} \quad s+L \leq t < \infty$$

when L is finite. This is of course the same thing as
defining

$$f(t) = 0 \quad \text{for} \quad 0 \leq t < s \quad \text{and for} \quad s+L < t < \infty;$$

indeed f = 0 does not contribute to the integral defining
F.

Thus, we can define F by the integral:

$$F(t) \ = \ \int_0^t f(x) \ dx \qquad \text{for } 0 \leq t < \infty$$

with

$$F(\infty) \ = \ \int_0^\infty f(x) \ dx \ = \ 1.$$

Clearly, this new definition preserves all properties of F
mentioned above. Thus, F(t) is a nondecreasing continuous
function of t, and the value F(t) represents the area
under the density curve, up to point t. Moreover, this
new definition of F has the advantage that it nicely takes
care of life times for which L is infinite. Then, F(t)
is less than 1 for all real t, and approaches 1 as
t → ∞.

As before, we have now

$$F(b) \ - \ F(a) \ = \ \int_a^b f(x) \ dx$$

and

$$F^c(t) \ = \ \int_t^\infty f(x) \ dx.$$

The following graph shows a typical d.f. and its density.

Average life. The d.f. F of the life time X provides
complete information about the behavior of X. However, it

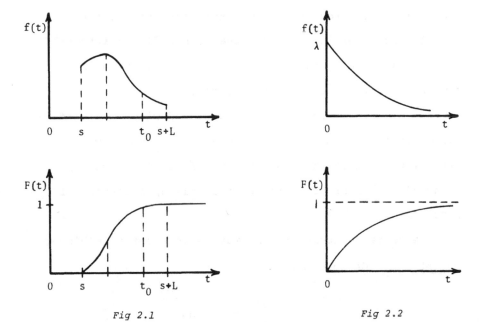

Fig 2.1   *Fig 2.2*

Typical distribution function F   *Exponential distribution.*
and density f.

is sufficient on many occasions to have only partial information about X. Thus instead of asking for the pr. of
waiting less than some value, it may be more informative to
know what the average waiting time would be.

We shall denote the <u>average life time</u> by μ or by
E(X) if dependence on X is stressed -- and define it by

$$\mu = \int_0^\infty tf(t)\ dt.$$

Here, one multiplies each value assumed by X with the
corresponding pr., say $t_n \cdot f(t_n) \cdot \Delta t_n$ for points t_n, and
sum over all n, in accordance with the notion of the
weighted average; the sum is replaced in the limit by the

integral. The words "the mean," the "expected life time"
are also used for μ. Observe that the defining integral
for μ is actually taken from s to $s+L$, when f
vanishes outside the interval $[s,s+L]$.

Note that μ is a number associated with X. Different densities f may produce the same μ, so knowing μ
tells little about X. Conversely, if f is known, the
value of μ is of great interest in all applications of
life times.

In certain situations one may be interested in higher
moments μ_n defined by:

$$\mu_n = \int_0^\infty t^n f(t)\ dt$$

where $n = 0,1,2,\ldots$. Observe that always $\mu_0 = 1$, and
$\mu_1 = \mu$. The center of gravity, and the moment of inertia
known from physics, are analogues of moments in probability.
We shall almost never use moments of order higher than 2.

Since the life time X is a variable quantity, one may
be interested in its fluctuation around the mean. A convenient way of measuring this fluctuation is with the help of
a variance, denoted by σ^2, and defined as

$$\sigma^2 = \mu_2 - \mu^2 = \int_0^\infty (t-\mu)^2 f(t)\ dt.$$

Observe that the variance is never negative. It can be verified that the variance is small when most of the pr. is
concentrated around the mean; that is the graph of density

f has most of its area around the mean. Conversely, if the graph is spread around the mean, then the variance is large. (The positive square root of the variance, denoted by σ, is called the standard deviation.)

As an exercise in integration, one can obtain from the defining integral the equivalent expression for the mean (which are sometimes useful in computations)

$$\mu \;=\; \int_0^\infty F^c(t)\ dt \qquad \text{(integrate by parts)}$$

With reference to the examples mentioned earlier one finds by simple integration:

Example 1 -- continued:

$$\mu_n \;=\; \frac{1}{L}\int_s^{s+L} t^n\ dt \;=\; \frac{1}{L}\left.\frac{t^{n+1}}{n+1}\right|_s^{s+L} \;=\; \frac{(s+L)^{n+1} - s^{n+1}}{L(n+1)} \;;$$

$$\mu \;=\; s + \frac{L}{2}\;;$$

$$\sigma^2 \;=\; \frac{L^2}{12}\;.$$

Example 2 -- continued:

$$\mu_n \;=\; \int_0^\infty t^n\lambda e^{-\lambda t}\ dt \;=\; \frac{1}{\lambda^n}\int_0^\infty x^n e^{-x}\ dx \;=\; \frac{n!}{\lambda^n}$$

$$\mu \;=\; \frac{1}{\lambda}\;,$$

$$\sigma^2 \;=\; \frac{1}{\lambda^2}\;.$$

Section 3:   Prolongation of Life Times

Suppose you wait in front of an occupied telephone
booth -- what is the pr. that the conversation which has been
in progress for some time, will end within the next couple
of minutes? You may ask what is the pr. that your car will
break down during the next month given that it performed
well since the last check up. If you treat a sick animal,
the natural question is to ask for the pr. of its survival,
when it is known that the animal is alive.

In these examples one actually considers the pr. that a
life time will be prolonged by an additional amount, given
that this life time has been continuing for some time. The
word "given" indicates that we calculate the pr. of some
event, given additional information -- in this case occurence
of another event. This is known as a conditional pr. If A
and B are two events, the event that A and B occur
simultaneously is denoted by A ∩ B. Its pr., written
pr(A ∩ B), could be taken as the pr. which interests us.
However, it is more convenient to work with the ratio of
pr(A ∩ B) to pr(B), where B is the conditioning event.
We thus define the conditional pr. of A given B by such
a ratio; in symbols, we shall write pr(A|B) -- which is
read "the pr. of A given B," the vertical line standing
for the word "given":

$$pr(A|B) \;=\; \frac{pr(A \cap B)}{pr(B)} \;.$$

(We must assume here that pr(B) is not equal to zero).

Thus pr(A|B) is the probability of the event A when it
is known that the event B has happened; in general this
pr. is different from pr(A) -- that event A happened (no
additional information).

   Let us return now to our life time X. We shall again
represent our events concerning X by appropriate intervals.
As the origin of the life time will not matter much in our
consierations, we shall assume for convenience that the
initial point s = 0; this will be our standing convention
from now on. As in Section 2, we shall use F for the d.f.
and f for the density of the life time X. Thus X
assumes values from 0 to L, inclusive, where L is the
duration of X (which may be infinite); if L is finite,
then $f(t) = 0$ and $F(t) = 1$ for $t \geq L$.

   Our task will be to find the pr. that the life time will
be prolonged by an additional amount, say h, when we know
that X lasted more than a specified value t. In terms of
events, (X > t+h) is the event that the life time is larger
than t + h; we shall take it as the event A in the above
definition. Similarly, (X > t) is the event that the life
time is larger than t; we take it for B. Thus, the pr.
we are going to find is:

$$pr(X > t+h \mid X > t).$$

In words, this is the pr. that the life time is continued
past t + h, given that it already lasted more than t. We
shall take this pr. as our defintiion of the pr. of prolongation by additional h; clearly $h \geq 0$ and $t \geq 0$.

Observe that $(X > t+h)$ implies $(X > t)$; indeed, if the life time lasted more than 15 years, it clearly lasted more than 10 years, but not conversely. Thus $(X > t+h) \cap (X > t) = (X > t+h)$. Substituting $A \cap B = (X > t+h)$ and $B = (X > t)$ in the above definition of the conditional pr.:

$$pr(X > t+h \mid X > t) \quad = \quad \frac{pr(X > t+h)}{pr(X > t)} \; .$$

Recall from Section 2 that in terms of the complementary d.f. F^c, these pr.'s are:

$$pr(X > t+h) \quad = \quad F^c(t+h), \qquad pr(X > t) \quad = \quad F^c(t).$$

Consequently, our pr. of prolongation by additional h is given by the formula:

$$pr(X > t+h \mid X > t) \quad = \quad \frac{F^c(t+h)}{F^c(t)} \; .$$

We shall now examine more closely this basic formula.

First of all observe that h can range from 0 to $L - t$, only (t is kept constant). For $h = 0$, the pr. of prolongation is 1, obviously; for $h = L - t$, the pr. of prolongation is evidently 0. Otherwise, the pr. decreases with increasing h -- this is also intuitive; indeed the longer prolongation of life time required, the smaller pr. of such an event. It should be noted that the pr. of prolongation depends in general on the instant t.

Examples: 1) For the case of uniform life (see Section 2):

$$f(t) = 1/L \quad \text{for} \quad 0 \le t \le L$$

and

$$F(t) = t/L \quad \text{for} \quad 0 \le t \le L.$$

Hence

$$F^c(t) = 1 - t/L \quad \text{and} \quad F^c(t+h) = 1 - (t+h)/L.$$

Consequently, the pr. of prolongation by h is:

$$F^c(t+h)/F^c(t) \quad = \quad \frac{L-t-h}{L-t} \quad = \quad 1 - h/(L-t)$$

for $0 \le h \le L-t$, and $0 \le t < L$.

2) For the case of exponential life (see Section 2):

$$f(t) \quad = \quad \lambda e^{-\lambda t} \quad \text{for} \quad 0 \le t$$

and

$$F(t) \quad = \quad 1 - e^{-\lambda t} \quad \text{for} \quad 0 \le t.$$

Hence

$$F^c(t) = e^{-\lambda t} \quad \text{and} \quad F^c(t+h) = e^{-\lambda(t+h)}.$$

Consequently, the pr. of prolongation by h is:

$$F^c(t+h)/F^c(t) \quad = \quad \frac{e^{-\lambda(t+h)}}{e^{-\lambda t}} \quad = \quad e^{-\lambda h}, \quad \text{for} \quad 0 \le h, \quad 0 \le t.$$

Observe that in this case the above pr. does not depend on t.   In other words, the pr. of prolongation by an additional amount does not depend on how long the life time has already lasted.   For example, the pr. that the telephone conversation will terminate during the next 5 minutes does not depend on how long it has already been in progress. This perhaps need not be exactly true in real life situations, but there are other life times for which this property holds, at least approximately.   Independence of the duration of a life time is the important property of the n.e.d.; it can be shown that the n.e.d. is the only distribution with this property.   We express this property by saying that the exponential distribution is memoryless; it does not remember how long it lasted, as far as prolongation is concerned.

Return now to the general case. Suppose that we would like to know the pr. of a complementary event that the life time will last till $t+h$, given it lasts more than t. This is clearly equal to 1 - pr. of prolongation:

$$\text{pr}(X \leq t+h \mid X > t) \;=\; 1 - F^c(t+h)/F^c(t) \;=\; \frac{F^c(t) - F^c(t+h)}{F^c(t)}$$

$$=\; \frac{F(t+h) - F(t)}{1 - F(t)}$$

for $0 \leq h \leq L - t$, and $0 \leq t < L$.

Observe that this pr. equals 0 for $h = 0$, and equals 1 for $h = L - t$. Moreover, for fixed t, this pr. increases with h.   Thus, as a function of h, it may be regarded as a <u>conditional</u> <u>d.f.</u> of a life time, given that the life time lasted more than t. We shall denote this conditional d.f. by K_t, the index t indicating dependence on t. Remember, however that t is kept fixed, and it is h that varies! It is essential that t be strictly less than L. Thus, recalling that F is the integral of f, the above expression for $K_t(h)$ can be written as:

$$K_t(h) \; = \; \frac{1}{F^c(t)} \int_t^{t+h} f(x)\,dx, \qquad \text{for } 0 \le h \le L-t, \quad 0 \le t < L$$

with $K_t(0) = 0$, $K_t(L-t) = 1$ (for fixed t). Note that for t = 0, one has $K_0(h) = F(h)$, and hence also $K_t(h) = 1$ for $h \ge L-t$.

Differentiation with respect to h (for fixed t) yields the conditional density $k_t(h) = dK_t(h)/dh$, which is given by

$$k_t(h) \; = \; \frac{f(t+h)}{F^c(t)}, \qquad \text{for } 0 \le h \le L-t.$$

Here again, $k_t(h) = 0$ for $h > L-t$.

Of special interest is the value of this density at h = 0, that is $k_t(0)$; it will be called the <u>hazard</u> <u>rate</u> of the life time, and will be denoted by r(t). Thus

$$r(t) \; = \; \frac{f(t)}{F^c(t)}, \qquad \text{for } 0 \le t < L,$$

with $r(t) = 0$ for $t > L$. Indeed $r(t)$ dh represents approximately the pr. that a t-old life time will fail

immediately during dh; r(t) is also called the <u>failure</u>
<u>rate</u>.

In problems concerning the reliability of equipment,
or components of equipment, one considers the <u>hazard func-</u>
<u>tion</u> R, defined by

$$R(t) = \int_0^t r(x)\ dx, \qquad \text{for } 0 \le t \le L.$$

The hazard rate r, as well as the hazard function R, determine uniquely the d.f. F and the density f of the
life time. It can be verified by a simple differentiation
that

$$F(t) = 1 - e^{-R(t)}, \qquad \text{for } 0 \le t \le L.$$

Consequently, in agreement with above:

$$f(t) = r(t)e^{-R(t)}.$$

Thus, if the hazard rate r(t) is given, then the life time
d.f. can be easily computed. Note however, that F(L) = 1
implies that R(L) is infinite, so also R(t) = ∞ for
t ≥ L (for finite L).

Examples: 1) (Continuation) For the uniform life,

$$K_t(h) = h/(L-t), \qquad \text{for } 0 \le h \le L-t$$

$$r(t) = 1/(L-t) \quad \text{and} \quad R(t) = -\log(1-t/L).$$

2) (Continuation) For the exponential life,

$$K_t(h) = 1 - e^{-\lambda h}, \quad \text{for} \quad 0 \leq h$$

and the hazard rate is constant

$$r(t) = \lambda$$

so $R(t) = \lambda t$, for $0 \leq t$.

Section 4: Bus Problem

Suppose that at a given stop buses are arriving independently of each other, the instants of their arrivals distributed along a time axis. We can take the <u>inter-arrival time</u> between two consecutive buses to be our life time X. Indeed, the time interval between two consecutive buses exhibits fluctuations which depend on chance, so X is a random variable. We shall assume that all inter-arrival times X have the same d.f. F with a density f and (finite) mean μ; thus, μ is the average interval between two consecutive buses.

The big question is now: what is the average waiting time of a passenger arriving at some random instant at the bus stop? The intuitive answer seems to be $\frac{1}{2} \mu$. Well, this is NOT the case, except when buses arrive at exactly fixed intervals of length μ. This is indeed a rather surprising result.

Once again: a passenger arrives at the bus stop
--naturally, there is no bus. The previous one has left a
long time ago and the next one will arrive (hopefully) soon.
The instant of the passenger's arrival is located at random
somewhere within the inter-arrival time (i.e. the life time)
X of buses. The passenger's <u>waiting</u> <u>time</u> W is of course
the time interval between his arrival and the arrival of the
next bus. Clearly, W is a random variable as its
fluctuations depend on chance. We can therefore consider
the event   (W ≤ t)   that the waiting time is t or less and
we shall write for its probability:

$$G(t) \;=\; \mathrm{pr}(W \le t).$$

In other words, the waiting time W is another life time,
and G is its distribution function. Our primary task will
be to find the form of G.

Despite its intuitive simplicity, the mathematical analysis of the problem is rather complicated, and cannot be
discussed here. Fortunately, the final result is surpris-
ingly simple, so with no hesitation we shall write down the
basic formula for G:

$$G(t) \;=\; \frac{1}{\mu} \int_0^t F^c(x)\, dx, \qquad \text{for } 0 \le t < \infty$$

Observe that G(t) increases with t, and that G(0) = 0,

$G(\infty) = 1$ as it should.   Moreover,   G   has density

$g = dG/dt$ given by:

$$g(t) \;=\; F^C(t)/\mu, \qquad \text{for} \quad 0 \le t < \infty$$

with $g(0) = 1/\mu$.

In general,   G   differs from   F.   One should also avoid confusing   G   with pr. of prolongation discussed in Section 3; the situations are entirely different.   Here we consider the remaining life time from a random instant till termination; thus   W   is sometimes called the residual life time.

We can now proceed to the evaluation of the average waiting time $E(W)$ which we simply denote by   w.   In accordance with what has been said about averaging (in Section 2), we must evaluate the integral

$$w \;=\; \int_0^\infty t\, g(t)\, dt.$$

Substituting for $g(t)$ and integrating by parts one has:

$$\int_0^\infty t\, g(t)\, dt \;=\; \frac{1}{\mu} \int_0^\infty t\, F^C(t)\, dt$$

$$=\; \frac{1}{\mu}\left[\, \frac{1}{2} t^2 F^C(t)\Big|_0^\infty + \frac{1}{2}\int_0^\infty t^2 f(t)\, dt \right]$$

$$=\; \frac{1}{2\mu}\int_0^\infty t^2 f(t)\, dt \;=\; \frac{1}{2\mu}\, \mu_2\, ,$$

where μ_2 is the second moment of   F

$$=\; \frac{1}{2\mu}(\sigma^2 + \mu^2),$$

because by definition $\sigma^2 = \mu_2 - \mu^2$. Consequently we have the final result:

$$\boxed{w = \frac{\mu}{2}(1 + \frac{\sigma^2}{\mu^2})}$$

Thus, w is always larger than $\frac{1}{2}\mu$, whenever σ^2 is not zero. The surprising fact is the dependence of the average waiting time w on the variance σ^2 of the life time distribution F. This in turn creates a paradox! For, if the d.f. F is such that $\sigma^2 > \mu^2$ (see Problem 5 for such an example), then clearly $w > \mu$, so the average waiting time is greater than the average inter-arrival time! It may even be infinite, if σ is infinite. Yet, as we defined it, the waiting time W is smaller than the inter-arrival time X.

Examples: 1) For uniform life time: As we have already seen:

$$F(t) = t/L, \quad \text{hence} \quad F^c(t) = 1 - t/L \quad \text{for} \quad 0 \le t \le L$$
$$= 1, \quad \text{hence} \quad = 0 \quad \text{for} \quad L \le t < \infty.$$

Moreover, $\mu = \frac{1}{2}L$, so

$$g(t) = (2/L)(1 - t/L) \quad \text{for} \quad 0 \le t \le L$$
$$= 0 \quad \text{for} \quad L \le t < \infty.$$

Furthermore, $\sigma^2 = L^2/12$, hence

$$w = L/3.$$

2) For exponential life: As we have already seen:

$$F(t) = 1 - e^{-\lambda t}, \quad \text{hence} \quad F^c(t) = e^{-\lambda t}, \quad \text{for} \quad 0 \le t < \infty.$$

Moreover, $\mu = 1/\lambda$, so

$$g(t) = \lambda e^{-\lambda t}, \quad \text{for} \quad 0 \le t < \infty.$$

Thus, we have another peculiarity of the n.e.d., namely that the densities f and g coincide. Indeed, the memoryless property of the exponential life implies that no matter when the passenger arrives, the distribution of the waiting time is exactly the same as if he would just miss the bus (and is forced to wait the whole inter-arrival time). The exponential life time does not pay any attention to late comers.

Since $\sigma^2 = 1/\lambda^2$, it is easy to check that $w = 1/\lambda = \mu$.

3) Constant life: This is a new example. Life times (inter-arrival times) are constant equal exactly to, say, L. No fluctuations are permitted (i.e. buses arrive at rigid fixed intervals. It is intuitively clear that now $\mu = L$, and as there are no fluctuations around the mean, the variance must be zero. Thus $\sigma^2 = 0$, hence $w = \frac{1}{2}\mu$. This is the intuitive answer mentioned at the beginning of this section.

But who ever saw buses arriving at regular intervals?

Section 5:   Combinations of Life Times

In many situations one must consider several life times simultaneously. If you have a device which is replaced when it breaks down, it may be of interest to know the total life span of several such replacements. In other situations, if several machines are in operation simultaneously, one may be interested in the shortest life time.

5.1 To handle such problems, information about joint behavior of life times is needed. This is provided by the joint density, or the joint d.f., of such life times. To be more specific, first consider two life times, say X and Y. The joint behavior of X and Y is described by their joint density f which is a function of two variables x and y. (For convenience, we shall used x and y as values of random variables X and Y, respectively). Clearly, $f(x,y) \geq 0$,   and

$$\int_0^\infty \int_0^\infty f(x,y) \ dx \ dy \ = \ 1.$$

We shall use our convention that f is 0 outside the range of values assumed by X and Y; restriction to nonnegative life times is nonessential.

Consider now the event, described jointly by X and Y, of the following form $(X \leq x, \ Y \leq y)$ -- this is the event that the first life time X is at most x, and the second life times Y is at most y. Write for the pr. of this event:

$$pr(X \le x, \; Y \le y) \;\; = \;\; F(x,y)$$

and call F the joint d.f. of X and Y.

In agreement with our standard procedure of defining probabilities by assigning numbers to events, we shall define the above pr. F(x,y) by the double integral of the density f:

$$F(x,y) \;\; = \;\; \int_0^x \int_0^y f(t,s) \; dt \; ds, \qquad 0 \le x < \infty,$$
$$0 \le y < \infty.$$

Note that if F is given, f can be obtained by partial differentiation:

$$f(x,y) \;\; = \;\; \partial^2 F(x,y)/\partial x \partial y.$$

Suppose that the joint density of X and Y is given, then the density f_1 of X, as well as the density f_2 of Y, can be obtained from f by integration:

$$f_1(x) \;\; = \;\; \int_0^\infty f(x,y) \; dy, \qquad f_2(y) \;\; = \;\; \int_0^\infty f(x,y) \; dx.$$

It is customary to call f_1 and f_2 the marginal densities. Hence, clearly the (marginal) d.f.'s of X and Y, respectively, are:

$$F_1(x) = \int_0^x f_1(t)\ dt, \qquad F_2(y) = \int_0^y f_2(s)\ ds.$$

The joint d.f. F, or the density f, describes the joint behavior of X and Y. From this joint behavior, as the above formulas show, we can deduce the individual behavior of X and of Y, separately. It just suffices to find the marginal densities (or d.f.'s). However, if we know only the individual behavior of X and Y, separately, nothing can be said in general about the joint behavior of X and Y; in other words, knowledge of marginal densities (d.f.'s) does not determine the joint density f (d.f. F).

There is, however, a very important special case when marginals determine the joint distribution. This case is referred to as <u>independence</u> of random variables X and Y, and is characterized by the property that the joint density f is simply the product of marginal densities f_1 and f_2:

$$f(x,y) = f_1(x) f_2(y), \qquad \text{for all } 0 \le x < \infty,$$
$$0 \le y < \infty.$$

This case introduces considerable mathematical simplifications. In the following we shall assume in most of our considerations that life times are independent. (We already did this in the previous section with independent inter-arrival times).

Simple (double) integration shows that independence can be characterized alternatively in terms of d.f.'s by factorization of the joint d.f. F:

$$F(x,y) \;=\; F_1(x)F_2(y), \quad \text{for all } 0 \le x < \infty, \;\; 0 \le y < \infty.$$

Conversely, using this relation as a definition of independence, simple differentiation yields the product of densities, as stated earlier.

The following two remarks may be useful, although we shall not make much use of them now.

Remark 1: The function $f(x,y)$ describes a surface in the 3-dimensional space, and the integral $F(x,y)$ represents the volume under that surface up to the point (x,y). The total volume under the surface is taken to be 1.

Remark 2: (On independence). Two events A and B are said to be <u>independent</u> if the pr. of their joint occurrence is the product of their individual pr.s.

$$\mathrm{pr}(A \cap B) \;=\; \mathrm{pr}(A)\,\mathrm{pr}(B).$$

Recall from Section 3, the definition of conditional pr.; independence implies that $\mathrm{pr}(A|B) = \mathrm{pr}(A)$, so A does not depend on B, in the probabilistic sense.

In agreement with our procedure of expressing events in terms of random variables, we can write:

$$A = (X \le x), \quad B = (Y \le y), \quad \text{so } A \cap B = (X \le x, \; Y \le y).$$

Hence, taking pr.s, one has from above:

$$pr(X \leq x, \ Y \leq y) \quad = \quad pr(X \leq x) \cdot pr(Y \leq y)$$

which is obviously

$$F(x,y) \quad = \quad F_1(x) F_2(y) .$$

<u>Extension</u>: Frequently we shall consider several life times simultaneously. It is convenient to denote them by $X_1, \ldots,$ X_n, where n is a positive integer. We shall also write for their d.f.'s and densities F_i and f_i, respectively, where $i = 1, \ldots, n$.

The joint behavior of these life times is described by their joint d.f.:

$$pr(X_1 \leq x_1, \ldots, X_n \leq x_n) = F(x_1, \ldots, x_n),$$

which is a function of n variables $x_1, \ldots, x_n$. This is the natural extension from the case n = 2, which we have just seen. Its probabilistic interpretation is only analogous, but analytically this becomes an n-directional integral. Thus, things become rather messy for n larger than 2, and we shall not consider such multi-dimensional distribution functions. Mathematicians, however, are not dicouraged by such difficulties. They introduce random vectors $X = (X_1, \ldots, X_n)$ and ordinary vectors for values $x = (x_1, \ldots, x_n)$ and write the above defining relation in the form resembling the one-dimensional case:

$$pr(X \leq x) = F(x) = \int_0^x f(t) \, dt,$$

which is the n-dimensional expression in disguise. Having
said that, we shall now ignore the fact.

However, there is a very important special case we
shall encounter very often, the case of independent life
times. We shall say that life times $X_1, \ldots, X_n$ are <u>independent</u> when their joint d.f. factorizes into the product of their
marginal d.f.'s:

$$F(x_1, \ldots, x_n) = F_1(x_1) \cdots F_n(x_n)$$

for <u>all</u> $x_1, \ldots, x_n$. Interpretation of this relation in terms
of events $(X_i \leq x_i)$ is the extension of the case n = 2.
Moreover, the above definition is equivalent to factorization of the joint density:

$$f(x_1, \ldots, x_n) = f_1(x_1) \cdots f_n(x_n).$$

It is this property that we are going to use when talking
about several independent life times.

****

5.2 When the joint density is given, one can evaluate pr.
of various events determined by life times X and Y, by
simple integration. The general rule is as follows. Sup-

pose that   A   is an event determined by   X   and   Y.   For
example,   (X+Y ≤ z),   or   (XY ≤ v),   etc.   These events   A
are in fact expressed by some function of   X   and   Y,   say
$Z = \varphi(X,Y)$.

Next, determine the region in the   x-y plane, corresponding to the event   A;   this again can be expressed in
terms of the function $\varphi(x,y)$.   Let   R(x,y)   be this region.
Then, the pr. of the event   A   is, in accordance with our
procedure of assigning numbers to events, given by

$$pr(A) \; = \; \iint_{R(x,y)} f(x,y) \; dx \; dy$$

The integral being evaluated over the region   R(x,y).   In
general, calculations are rather involved, so we shall
restrict ourselves to a few simple cases of interest for us.
Our standing assumption is now that   X   and   Y   are independent.

$$*****************$$

Sum of two life times: Let   X   and   Y   be two independent
life times, with densities f_1 and f_2,   respectively.
The total life time is   Z = X + Y.   It is required to find
the density   g   of   Z.

The answer is given by the following integrals, known
as convolution of densities:

$$g(z) \; = \; \int_0^z f_1(x) f_2(z-x) \; dx \; = \; \int_0^z f_1(z-y) f_2(y) \; dy, \quad 0 \le z < \infty$$

This can be justified as follows. For the event
$A = (X+Y \leq z)$, the corresponding region is $R(x,y) = (x+y \leq z)$,
so according to the above procedure:

$$\text{pr}(X+Y \leq z) = \iint_{x+y \leq z} f_1(x) f_2(y) \ dx \ dy$$

$$= \int_0^z f_1(x) \ dx \int_0^{z-x} f_2(y) \ dy$$

using independence, and integrating first along the y-axis,
then along the x-axis. The above integral gives, of course,
the d.f. $G(z)$ of Z. Hence, by differentiation $g(z) = dG/dz$ one obtains the first expression for g. The second
follows in the same manner, by integrating first with
respect to x, and then with respect to y. Note that here
the region $(x+y \leq z)$ is a triangle in the first quadrant,
bounded by axes and the line $x + y = z$.

Computing the average total life time $E(Z)$ it can be
verified that

$$E(X+Y) = E(X) + E(Y).$$

So average life times add. (It can be shown that this property holds in general, irrespective of whether life times
are independent.)

Note: For independent life times X and Y one has:

$$\text{var}(X+Y) = \text{var}(X) + \text{var}(Y).$$

This relation is NOT true in general for dependent life
times.

****

In later sections we shall consider other combinations
of life times. At the moment, let us stop for some examples.

****

<u>Example 1</u>: Suppose that X and Y are independent and
have the same n.e.d.:

$$f_1(x) = \lambda e^{-\lambda x}, \quad 0 \le x < \infty, \qquad f_2(y) = \lambda e^{-\lambda y}, \quad 0 \le y < \infty.$$

Hence, the density of X + Y is:

$$g(z) = \int_0^z \lambda e^{-\lambda x} \, \lambda e^{-\lambda(z-x)} \, dx = e^{-\lambda z} \lambda^2 \int_0^z dx = (\lambda z) e^{-\lambda z} \lambda$$

for $0 \le z < \infty$, and

$$E(Z) = \int_0^\infty z g(z) \, dz = \frac{2}{\lambda}.$$

<u>Example 2</u>: Suppose that X and Y are independent and
have the same uniform density:

$$f_1(x) = \frac{1}{L}, \qquad 0 \le x \le L \qquad\qquad f_2(y) = \frac{1}{L}, \qquad 0 \le y \le L$$
$$ = 0, \qquad x > L \qquad\qquad\qquad = 0, \qquad y > L.$$

Hence, the density of X + Y is

$$g(z) = \int \frac{1}{L} \cdot \frac{1}{L} \, dx$$

where the integral is taken over the region such that
simultaneously: $0 \le x \le L$, $0 \le z-x \le L$.

One must consider separately two regions (where always $0 \leq z \leq 2L$)

$$0 \leq z \leq L \qquad \text{for which} \qquad 0 \leq x \leq z$$
$$L \leq z \leq 2L \qquad \text{for which} \qquad z-L \leq x \leq L$$

so

$$g(z) \;=\; \int_0^z \frac{dx}{L^2} \;=\; \frac{z}{L^2} \qquad \text{for} \quad 0 \leq z \leq L$$

$$g(z) \;=\; \int_{z-L}^L \frac{dx}{L^2} \;=\; \frac{2L-z}{L^2} \qquad \text{for} \quad L \leq z < 2L$$

$$g(z) \;=\; 0 \qquad \qquad \text{for} \quad z > 2L.$$

The graph of $g(z)$ is a triangle with vertices at points $(0,0)$, $(L, \frac{1}{L})$ and $(2L, 0)$.

Note: $E(X+Y) = 2 \cdot \frac{L}{2} = L.$

****

<u>Total life time</u>. There is a golden rule that says that if you can do something once, you can do it several times. This is the case when you have several life times $X_1, \ldots, X_n$; where n is a fixed integer, and wish to consider the total life time:

$$S_n = X_1 + \cdots + X_n, \qquad n \geq 1.$$

For example, if each X represents time needed to read a book, then S_n will be time for reading n books.

Assume now that $X_1, \ldots, X_n$ are independent, and let

$f_1, \ldots, f_n$ be their densities, respectively. Denote by g_n
the density of the total life time S_n. The most natural way
of finding g_n is by reducing the problem to the case of two
life times. Indeed, write

$$S_n = S_{n-1} + X_n, \qquad n \geq 2.$$

It is intuitively clear that life times S_{n-1} and X_n are independent. Hence, by the previous reasoning, we have immediately

$$g_n(z) = \int_0^z g_{n-1}(z - x) f_n(x) dx, \qquad 0 \leq z < \infty, \; n \geq 2.$$

This is a recurrence relation—it means that calculations of
g_n must be done step by step. Starting with $g_1 = f_1$, compute g_2. Using this g_2 and f_3, compute g_3, and so on. Unfortunately, such calculations are very tedious and in most
cases prohibitive, even in the simplifying situation when
all life times X have the same density (as in the case
treated in Section 16). In a few lucky cases, however, close
explicit formulae for g_n may be obtained (as in the exponential case treated in Section 16).

Nevertheless, the average total life time is easily
found to be

$$\mathbf{E}S_n = \mathbf{E}X_1 + \cdots + \mathbf{E}X_n$$

(irrespective of individual X's being independent or not.

However, for independent life times X's, var S_n is the sum of individual variances, but this is not true for dependent life times in general.

Section 6: Extreme Life Times

Suppose that in an experiment in Psychology there are n subjects who must perform a certain task (say, children working on a puzzle). The experiment is arranged so that all n subjects start simultaneously, at instant 0 say, and the objective is to find the shortest (as well as the longest) time needed to perform the task. Of course, it is not known at the beginning which of the subjects will finish first, and which will finish last. It is therefore required to find the distribution of the shortest time, and of the longest time, needed (irrespective of the individual subjects).

Represent by $X_1, X_2, \ldots, X_n$ the length of time (life time) needed to perform the task by subjects $1, 2, \ldots, n$, respectively. Assume that life times $X_1, X_2, \ldots, X_n$ are independent, and denote by $f_1, f_2, \ldots, f_n$ their densities, and by $F_1, F_2, \ldots, F_n$ their d.f.'s.

The waiting time for the <u>first completion</u> is the smallest life time out of $X_1, X_2, \ldots, X_n$. Denote it by M_-, so

$$M_- = \min(X_1, X_2, \ldots, X_n).$$

Clearly, M_- is a life time, and write G_- for its d.f.:

$$G_-(t) = \text{pr}(M_- \leq t), \qquad 0 \leq t < \infty.$$

The event $(M_- > t)$ is the simultaneous realization of the n events $(X_1 > t), (X_2 > t), \ldots, (X_n > t)$ whose pr.'s are, respectively, $F_1^C(t), F_2^C(t), \ldots, F_n^C(t)$. Because of the assumed independence (see Section 5) the pr.'s multiply so

$$pr(M_- > t) = pr(X_1 > t) pr(X_2 > t) \ldots pr(X_n > t).$$

In other words, the complementary d.f. is:

$$G_-^C(t) = F_1^C(t) F_2^C(t) \ldots F_n^C(t)$$

and therefore $G_-(t) = 1 - G_-^C(t)$, for $0 \leq t < \infty$. The corresponding density is found by differentiation: $g_-(t) = dG_-(t)/dt$.

In most cases of interest, the life times are identically distributed with the common d.f. F with density f. Consequently, the above expression simplifies to:

$$G_-(t) = 1 - [1 - F(t)]^n$$

and the corresponding density is

$$g_-(t) = n[1 - F(t)]^{n-1} f(t).$$

****

The waiting time for the last completion is the largest life time out of $X_1, X_2, \ldots, X_n$. Denote it by M_+, so

$$M_+ = \max(X_1, X_2, \ldots, X_n).$$

Clearly, M_+ is a life time, and write G_+ for its d.f.:

$$G_+(t) \quad = \quad pr(M_+ \le t), \qquad\qquad 0 \le t < \infty.$$

The event $(M_+ \le t)$ is the simultaneous realization of the n events $(X_1 \le t), (X_2 \le t), \ldots, (X_n \le t)$ whose pr.'s are, respectively, $F_1(t), F_2(t), \ldots, F_n(t)$. Because of the assumed independence (see Section 5) the pr.'s multiply so

$$pr(M_+ \le t) \quad = \quad pr(X_1 \le t)\, pr(X_2 \le t) \ \ldots \ pr(X_n \le t).$$

In other words, the d.f. is:

$$G_+(t) \quad = \quad F_1(t) F_2(t) \ \ldots \ F_n(t), \qquad 0 \le t < \infty.$$

The corresponding density is found by differentiation $g_+(t) = dG_+(t)/dt$.

When the life times are identically distributed with the same F and f, then

$$G_+(t) \quad = \quad [F(t)]^n$$

and the corresponding density is:

$$g_+(t) \quad = \quad n[F(t)]^{n-1} f(t) \quad .$$

When distributions of extreme life times M_- and M_+ are known, one can compute their mean values $E(M_-)$ and

$E(M_+)$, as well as variances, using the method discussed in Section 2.

****

Example: Exponential life. All life times have the same n.e.d. (for $0 \leq t < \infty$):

$$F(t) = 1 - e^{-\lambda t}$$

$$f(t) = \lambda e^{-\lambda t}$$

$$\mu = 1/\lambda.$$

Hence

$$g_-(t) = n\lambda e^{-n\lambda t}.$$

This is again the n.e.d., but with a parameter $n\lambda$, so

$$E(M_-) = 1/(n\lambda).$$

On the other hand

$$g_+(t) = n\lambda e^{-\lambda t}(1-e^{-\lambda t})^{n-1}, \qquad G_+(t) = (1-e^{-\lambda t})^n.$$

In order to compute the mean $E(M_+)$, it is more convenient to use the formula involving $G_+^c(t)$, as noted in Section 2 and in Section 4:

$$E(M_+) = \int_0^\infty G_+^c(t)\ dt = \int_0^\infty [1 - (1-e^{-\lambda t})^n]\ dt$$

$$= \int_0^1 (1-z^n)\ \frac{dz}{\lambda(1-z)}\ ,$$

after substitution of $1 - e^{-\lambda t} = z$

$$\lambda e^{-\lambda t}\, dt = dz$$

$$\lambda (1-z) dt = dz$$

$$= \frac{1}{\lambda} \int_0^1 (1+z+z^2+\ldots+z^{n-1})\, dz,$$

because $(1-z)(1+z+\ldots+z^{n-1}) = 1 - z^n$

$$= \frac{1}{\lambda} (1 + \frac{1}{2} + \frac{1}{3} + \ldots + \frac{1}{n}) \quad .$$

Note that for $n \neq 1$:

$$E(M_-) < \frac{1}{\lambda} < E(M_+),$$

and that $\lim_{n\to\infty} E(M_-) = 0, \quad \lim_{n\to\infty} E(M_+) = \infty.$

****

The following practical problem is treated in exactly the same manner, stressing once more the fact that problems from different fields maybe mathematically identical.

Consider a system of n independent components, arranged either in series or in parallel (see figures) in which each component functions for a random length of time having a d.f. F, the same for all components, with density f. Thus, $F^c(t) = 1 - F(t)$ is the pr. that a component is functioning at time t (i.e., its life time is greater than t).

a) In series arrangement, the system functions when <u>all</u> its components are functioning.

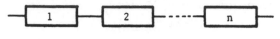

Fig 6.1

Series arrangement.

Thus, the d.f. G_s of the amount of time that the system functions is given by:

$$1 - G_s(t) \;=\; [1 - F(t)]^n.$$

b) In parallel arrangement, the system functions when <u>at</u> <u>least</u> <u>one</u> of its components is functioning.

Thus, the d.f. G_p of the amount of time that the system is functioning is given by:

$$1 - G_p(t) = 1 - [F(t)]^n.$$

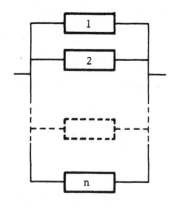

Fig 6.2

Parallel arrangement.

Note: expressions in (a) and (b) are known as <u>reliability</u> of a system.

****

<u>Example</u>: Suppose now that life times of the components are uniform: $F(t) = t/L$ (for $0 \leq t \leq L$). We have now for $0 \leq t \leq L$:

(i) for series system:

$$1 - G_s(t) = (1 - \frac{t}{L})^n , \qquad g_s(t) = \frac{n}{L}(1 - \frac{t}{L})^{n-1} .$$

(ii) for the parallel system:

$$G_p(t) = (\frac{t}{L})^n , \qquad g_p(t) = \frac{n}{L}(\frac{t}{L})^{n-1} .$$

Denote by M_s and M_p the life time when the system functions, in series and in parallel arrangement, respectively. These mean life times are given by

$$E(M_s) = \frac{L}{n+1} , \qquad E(M_p) = \frac{n}{n+1} L.$$

(Recall that the mean life of each component is $L/2$.) Note that (for fixed L), when the number n of components becomes very large, then $E(M_s)$ tends to zero, but $E(M_p)$ tends to L.

Section 7: Great Expectations

Frequently, direct observation of life time may be too difficult to perform, or one may be more interested in some

other aspects of this life time. For example, recording
apparatus plots a graph of life time, subject to change of
scale; in design of a facility cost of life (waiting time)
may be of importance. Thus, the actual life time is
replaced by a secondary quantity which is some kind of a
function of the life time. For convenience, we shall call
it a cost function. Its values may be expressed in dollars,
or in some other units. In Economics, a term utility is
also used for the same purpose; we can also speak about gain
or loss, associated with life times. The term cost function
will embrace all such applications.

It may happen that in some situation, the pr. of a
long life time exceeding a fixed duration may be comfortably
large, but cost may be very large; to reduce costs, shorter
life times may be preferred. The most common method of
assessing costs is to evaluate the average cost of the life
time. We shall now discuss various aspects of such evalua-
tions.

Let X be a life time with density f, d.f. F and
mean life time μ. We shall denote a cost function by φ.
Thus, φ(X) is a random variable representing cost associated with a life time X; in particular, if life time X
assumes the value x, then the cost function assumes the
value φ(x). Observe that φ(X) may assume values of both
signs (positive values representing gain, negative loss --
or conversely, depending on the point of view).

The average cost is now defined by:

$$E\varphi(X) \;=\; \int_0^\infty \varphi(x) f(x) \; dx \; .$$

Note that for $\varphi(x) = x$ the integral yields the usual mean μ, whereas for $\varphi(x) = x^n$ one obtains the moments μ_n (see Section 2). The point is that now we shall evaluate this integral for arbitrary functions φ.

Since $\varphi(X) = Y$ is a random variable, one could be interested in its distribution, and then evaluate $E(Y)$ as in Section 2; this of course can be done (using some calculus), but for evaluation of expectation it is, fortunately unecessary. Perhaps it should be added that although the average cost has been defined above for a life time X, obviously the same definition applies to other life times, like the extreme life times (from Section 6), residual life times (from Section 4), and others, provided that f is now interpreted as an appropriate density.

Example 1: Linear Cost. Suppose that life is measured in days, and let a be the rate in dollars per day, and b the cost in dollars of initial operation. Thus, total cost for x days is (in dollars):

$$\varphi(x) = ax + b, \quad (x \geq 0).$$

The average cost is:

$$E\,\varphi(X) = \int_0^\infty (ax+b)\,f(x)\,dx = a\mu + b$$

irrespective of the form of density f.

Example 2: Quadratic cost. Suppose that m is a desirable life of equipment in some construction project. Then, X - m is a random fluctuation around m of the life time X. To have positive values, it is convenient to consider rather $(X-m)^2$ as the "error," departure from m. If c is the unit cost, then the total cost of a discrepancy is $c(X-m)^2$. The average cost is

$$Ec(X-m)^2 = c \int_0^\infty (x-m)^2 f(x) \ dx = c[\sigma^2 + (\mu-m)^2]$$

where $\sigma^2 = \text{var}(X)$. The integral is simply evaluated in the following manner:

$$\int_0^\infty (x-m)^2 f(x) \ dx = \int_0^\infty [(x-\mu) + (\mu-m)]^2 f(x) \ dx$$

$$= \sigma^2 + 2(\mu-m) \int_0^\infty (x-\mu) f(x) \ dx + (\mu-m)^2$$

$$= \sigma^2 + (\mu-m)^2.$$

Suppose now that we wish to minimize the average cost by selecting appropriate m. From the above expression for the expected cost it is clearly seen that minimum will be achieved when $m = \mu$. Thus, the mean life time minimizes the quadratic cost, not matter what form of f.

Example 3: Kinetic energy of a particle of mass m is given by $\frac{1}{2} mx^2$, where x is its velocity. Taking vel-

ocity as a life time X, the average kinetic energy is
given by

$$E \frac{1}{2} mX^2 \;=\; \frac{1}{2} m \int_0^\infty x^2 f(x) \; dx \;=\; \frac{1}{2} m \mu_2$$

where μ_2 is the second moment.

Note that if X is the radius of a circle, the same
argument gives for the average area $\pi\mu_2$ (which is more than
$\pi\mu^2$, where μ is the average radius).

Example 4: Cut-off point. Suppose that a facility operates
in such a way that if the waiting time X is less than a
fixed amount a, then no loss is incurred. If the waiting
time exceeds a, then the fixed penalty p is paid. The
cost function is now:

$$\varphi(x) \;=\; 0 \qquad \text{for} \quad x < a$$
$$\;=\; p \qquad \text{for} \quad x \geq a.$$

The average cost is now:

$$E\,\varphi(X) \;=\; \int_0^a 0\,f(x) \; dx + p \int_a^\infty f(x) \; dx \;=\; pF^c(a).$$

Suppose now that the penalty p is proportional to the
cut-off point a, so p = ca, and that X has exponential life. Hence the average cost is

$$cae^{-\lambda a},$$

and it is clear that the maximal average cost will occur at the cut-off point $a = \mu = 1/\lambda$, and equals $c\mu/e$.

Example 5: A signal produced by a device at time x is of the form $\sin \omega x$, where ω is the angular frequency. Hence, the average value of the signal for exponential life time X is:

$$E \sin \omega X = \int_0^\infty \lambda e^{-\lambda x} \sin \omega x \, dx = \frac{\lambda \omega}{\lambda^2 + \omega^2}$$

(the integral being evaluated by parts, twice).

Example 6: Strange life time. Suppose that density is given by the form:

$$f(x) = \frac{2}{\pi (1+x^2)}, \quad 0 \le x < \infty.$$

It is easy to verify that

$$\int_0^\infty f(x) \, dx = 1.$$

Let's look at its mean value:

$$\mu = \frac{2}{\pi} \int_0^\infty \frac{x}{1+x^2} \, dx = \frac{1}{\pi} \int_0^\infty \frac{dy}{1+y} = \frac{1}{\pi} \log(1+y) \bigg|_0^\infty = \infty.$$

The mean is infinite! This may be a little puzzling, but intuitively it means that the average value is rather large, larger than any number. To cut it down to size, select a suitable cost function. Indeed, $\varphi(x) = \arctan x$ will do:

$$E(\arctan X) = \frac{2}{\pi} \int_0^\infty \arctan x \frac{dx}{1+x^2} = \frac{2}{\pi} \int_0^{\pi/2} y \, dy$$

$$= \frac{2}{\pi} \frac{1}{2} y^2 \Big|_0^{\pi/2} = \frac{\pi}{4} .$$

Example 7: One must be careful to avoid confusing $E(1/X)$ with $1/E(X)$. Suppose that X is exponential life, so

$$E(1/X) = \lambda \int_0^\infty \frac{1}{x} e^{-\lambda x} \, dx = \lambda \int_0^\infty \frac{1}{y} e^{-y} \, dy = \infty$$

(the integral is known to diverge). Similarly, for uniform life $f(x) = 1/L$ for $s \le x \le s+L$:

$$E(1/X) = \frac{1}{L} \int_0^{s+L} \frac{1}{x} \, dx = \frac{1}{L} \log x \Big|_s^{s+L} = \frac{1}{L} \log \frac{s+L}{s}.$$

and this is finite for $s > 0$, and infinite for $s = 0$.

Yet, in both cases $1/E(X) = 1/\mu$ which is finite.

Example 8: It is obvious that if the cost function is $\varphi(x) = x - \mu$, then the average cost is always zero, for any density f:

$$E(X-\mu) = \int_0^\infty (x-\mu) f(x) \, dx = \mu - \mu = 0.$$

****

When two life times X and Y are considered simultaneously, the cost function φ depends now on two vari-

ables, so $\varphi(X,Y)$. The average cost is then defined by the
double integral:

$$E\varphi(X,Y) = \int_0^\infty \int_0^\infty \varphi(x,y) f(x,y) \, dx \, dy$$

where $f(x,y)$ is the joint density, as introduced in Section 5. Recall that in the important case when life times
X and Y are independent, then $f(x,y) = f_1(x) f_2(y)$,
where f_1 and f_2 are marginal densities of X and of Y,
respectively (see Section 5).

Thus in the case of independence:

$$E\varphi(X,Y) = \int_0^\infty \int_0^\infty \varphi(x,y) f_1(x) f_2(y) \, dx \, dy.$$

Example 9: Linear cost. This is the extension of Example
1, with $\varphi(x,y) = ax + by + c$, where a and b are rates,
and c is the initial cost. The average cost is:

$$E(aX+bY+c) = \int_0^\infty \int_0^\infty (ax+by+c) f(x,y) \, dx \, dy$$

$$= \int_0^\infty axf_1(x) \, dx + \int_0^\infty byf_2(y) \, dy + c$$

$$= aE(X) + bE(Y) + c$$

irrespective of the form of the joint density.

Example 10: Product cost. Suppose that cost is proportional to the product XY of life times, and that X and

Y are independent. Thus, $\varphi(X,Y) = cXY$, and the average cost is

$$E(cXY) \;=\; c \int_0^\infty \int_0^\infty xy f_1(x) f_2(y) \; dx \; dy$$

$$= \; c \int_0^\infty x f_1(x) \; dx \cdot \int_0^\infty y f_2(y) \; dy \;=\; cE(X)E(Y).$$

(This formula does not hold, in general, for dependent life times).

Example 11: Distance of a random point from the origin satisfies the relation $R^2 = X^2 + Y^2$, where (X,Y) are coordinates of a point in the first quadrant. The area of (a quarter) of a circle with radius R is $A = (\pi/4)R^2$. The average area is therefore

$$E(A) \;=\; \frac{1}{4}\pi \int_0^\infty \int_0^\infty (x^2+y^2) f(x,y) \; dx \; dy \;=\; \frac{1}{4}\pi \, [E(X^2) + E(Y^2)]$$

and this again depends on second moments (and not on square of the first moment); see Example 3.

Example 12: What is the average distance between two points chosen at random on the unit interval? To formulate this problem a little more precisely, suppose that the position of the points are represented by X and Y; the distance between them is $|X-Y|$, the absolute value is taken because distance is positive. Furthermore, we assume that X and Y are independent life times, each with a uniform

distribution on the interval from 0 to 1 (i.e., L = 1).
Thus, the average distance is:

$$E|X - Y| \quad = \quad \int_0^1 \int_0^1 |x - y| \ dx \ dy \quad = \quad \frac{1}{3} \ .$$

Example 13: The analogous problem to Example 12 with n.e.d.
on the positive half-axis leads to

$$E|X - Y| \quad = \quad \int_0^\infty \int_0^\infty |x - y| \ \lambda e^{-\lambda x} \ \lambda e^{-\lambda y} \ dx \ dy \quad = \quad \frac{1}{\lambda} \ .$$

The integral is rather tedious to evaluate, but the final
result is of interest: the average distance is exactly
equal to the average life.

Example 14: In contrast with Example 12 and Example 13, if
cost is expressed as $(X - Y)^2$, the evaluation of average
cost for independent X and Y is very easy:

$$E(X-Y)^2 \quad = \quad \int_0^\infty \int_0^\infty (x-y)^2 f_1(x) f_2(y) \ dx \ dy$$

$$= \quad var(X) + var(Y) + (EX - EY)^2$$

irrespective of the form of densities.

Note on the Laplace transform

We have already seen that the exponential function has
several special properties. It is therefore no surprise

that the exponential cost function lives up to its name:

$$\varphi(x) = e^{-\alpha x}, \quad \text{for} \quad x \geq 0$$

where $\alpha \geq 0$ is a parameter. Physically, this cost function represents "declining costs" (or discount) at the rate
α. The more important, however, is the average cost

$$E\varphi(X) = Ee^{-\alpha X}$$

of a non-negative life time X having density f. This
average cost regarded as a function of α has rather
remarkable properties, quite apart from its interpretation
as a cost. Indeed, this new function, to be denoted by f^*
and defined by:

$$f^*(\alpha) = \int_0^\infty e^{-\alpha x} f(x)\, dx \quad \text{for} \quad \alpha \geq 0$$

occurs frequently in applications, and deserves a special
name; it is called the <u>Laplace</u> <u>transform</u> of the density f.
Its primary usefulness is as a computational tool.

First of all note that no matter what form of the
density f, its transform f^* is always a continuous function of α. Moreover, $f^*(0) = 1$.

Differentiation yields:

$$df^*(\alpha)/d\alpha = -\int_0^\infty x e^{-\alpha x} f(x)\, dx .$$

Laplace transforms are very useful in evaluation of properties of some cost functions or combination of life times, provided of course that resulting expressions are easier to work with than those obtained by direct methods.

For example, if Y is a linear transformation of a life time X, that is Y = aX + b, then the transform of the density of Y can be expressed in terms of the transform of density of X as follows:

$$Ee^{-\alpha Y} = Ee^{-\alpha(aX+b)} = e^{-\alpha b}Ee^{-\alpha aX} = e^{-\alpha b}f^*(\alpha a).$$

From this moments of Y can be easily computed.

The following expressions (stated here without proof) are very useful:

$$\lim_{\alpha \to \infty} f^*(\alpha) = 0, \qquad \lim_{\alpha \to \infty} \alpha f^*(\alpha) = \lim_{t \to 0} f(t),$$

$$\lim_{\alpha \to 0} \alpha f^*(\alpha) = \lim_{t \to \infty} f(t).$$

Another illustration of advantages offered by Laplace transforms is provided by the transform of a convolution integral. Recall from Section 5 that if X and Y are two independent life times with densities f_1 and f_2 respectively, then the total life time Z = X + Y has a density g given by convolution integral

$$g(z) = \int_0^z f_1(x)f_2(z-x)\ dx.$$

Hence, the value of the first derivative of f^* at $\alpha = 0$ yields the (negative of) the first moment of X:

$$df^*(\alpha)/d\alpha \Big|_{\alpha=0} = -E(X).$$

Similarly, the second derivative:

$$d^2f^*(\alpha)/d\alpha^2 = \int_0^\infty x^2 e^{-\alpha x} f(x) \ dx$$

yields the second moment:

$$d^2f^*(\alpha)/d\alpha^2 \Big|_{\alpha=0} = E(X^2).$$

More generally, the n-th differentiation yields:

$$d^nf^*(\alpha)/d\alpha^n \Big|_{\alpha=0} = (-1)^n E(X^n).$$

Thus, one can compute moments of X by differentiation -- and this is frequently very useful.

Using the expansion:

$$e^{-\alpha x} = \sum_{n=0}^\infty (-1)^n (\alpha x)^n / n!$$

it follows that, if all moments exist then:

$$f^*(\alpha) = \sum_{n=0}^\infty (-1)^n \alpha^n \mu_n / n!$$

so moments can be obtained from expansion of f^* in powers of α.

Furthermore, it is easy to check (by double integration) that if F is a d.f. with density f, then the Laplace transforms of F and F^C are, respectively:

$$\int_0^\infty e^{-\alpha x} F(x) \, dx = f^*(\alpha)/\alpha,$$

$$\int_0^\infty e^{-\alpha x} F^C(x) \, dx = [1-f^*(\alpha)]/\alpha .$$

Similarly, the Laplace transform of the derivative $f'(x) = df(x)/dx$ is

$$\int_0^\infty e^{-\alpha x} f'(x) \, dx = \alpha f^*(\alpha) - f(0).$$

Examples:

  i) Uniform life:

$$f^*(\alpha) = \frac{1}{L} \int_s^{s+L} e^{-\alpha x} \, dx = \frac{e^{-\alpha s}(1-e^{-\alpha L})}{L\alpha} .$$

  ii) Exponential life:

$$f^*(\alpha) = \int_0^\infty e^{-\alpha x} \lambda e^{-\lambda x} \, dx = \frac{\lambda}{\lambda+\alpha} .$$

  iii) Constant life (see Section 4):

$$f^*(\alpha) = e^{-\alpha L}.$$

Laplace transforms are very useful in evaluation of properties of some cost functions or combination of life times, provided of course that resulting expressions are easier to work with than those obtained by direct methods.

For example, if Y is a linear transformation of a life time X, that is $Y = aX + b$, then the transform of the density of Y can be expressed in terms of the transform of density of X as follows:

$$Ee^{-\alpha Y} = Ee^{-\alpha(aX+b)} = e^{-\alpha b}Ee^{-\alpha aX} = e^{-\alpha b}f^*(\alpha a).$$

From this moments of Y can be easily computed.

The following expressions (stated here without proof) are very useful:

$$\lim_{\alpha \to \infty} f^*(\alpha) = 0, \qquad \lim_{\alpha \to \infty} \alpha f^*(\alpha) = \lim_{t \to 0} f(t),$$

$$\lim_{\alpha \to 0} \alpha f^*(\alpha) = \lim_{t \to \infty} f(t).$$

Another illustration of advantages offered by Laplace transforms is provided by the transform of a convolution integral. Recall from Section 5 that if X and Y are two independent life times with densities f_1 and f_2 respectively, then the total life time $Z = X + Y$ has a density g given by convolution integral

$$g(z) = \int_0^z f_1(x)f_2(z-x) \, dx.$$

In the present case, the Laplace transform of g is given simply by the product of the Laplace transforms of f_1 and of f_2, that is:

$$g^*(\alpha) \quad = \quad f_1^*(\alpha) f_2^*(\alpha).$$

This is easily justified as in Example 10 above, making use of independence:

$$Ee^{-\alpha Z} \quad = \quad Ee^{-\alpha(X+Y)} \quad = \quad E(e^{-\alpha X}e^{-\alpha Y}) \quad = \quad Ee^{-\alpha X} \cdot Ee^{-\alpha Y}.$$

For those who like double integrals, the same result may be obtained alternatively by direct evaluation of the integral, by using the explicit form of density g.   Thus:

$$g^*(\alpha) \quad = \quad \int_0^\infty e^{-\alpha z} g(z) \; dz \quad = \quad \int_0^\infty e^{-\alpha z} \; dz \int_0^z f_1(x) f_2(z-x) \; dx$$

$$= \quad \int_0^\infty f_1(x) e^{-\alpha x} \; dx \int_x^\infty e^{-\alpha(z-x)} f_2(z-x) \; dz$$

$$= \quad f_1^*(\alpha) f_2^*(\alpha).$$

As an example, consider Example 1 in Section 5.   The Laplace transform is simply

$$g^*(\alpha) \quad = \quad (\frac{\lambda}{\lambda+\alpha})^2 \; .$$

****

<u>Note</u>. Consider the total life time (as in Section 5):

$$S_n = X_1 + \cdots + X_n$$

of n <u>independent</u> life times $X_1, \ldots, X_n$, with Laplace transforms $f_1^*(\alpha), \ldots, f_n^*(\alpha)$. By the same argument, the Laplace
transform of the density g_n is found to be

$$g_n^*(\alpha) = f_1^*(\alpha) \cdots f_n^*(\alpha).$$

In particular, when $X_1, \ldots, X_n$ have common density f, then

$$g_n^*(\alpha) = [f^*(\alpha)]^n.$$

Thus, getting $g_n^*(\alpha)$ is easy, but to obtain from it the explicit form of $g_n(t)$ may be rather cumbersome, to put it
mildly. Fortunately, in some special cases this can be
done. We shall see a few of them later.

For further applications of Laplace transforms see
Sections 17 and 24.

Section 8: Double Scotch

Recall that in Section 5 we have defined the pr. of an
event determined by two random variables X and Y as an
integral of their joint density f over a set R in the
plane where X and Y take their values:

$$\text{pr}(A) = \iint_{R(x,y)} f(x,y) \, dx \, dy.$$

In other words, with each set R we associate a real number, the value of the above integral as the pr. of the
corresponding event A.

The integral to be evaluated is a double integral, and
in this section some illustrative examples will be worked
out.

1. <u>Two friends meeting</u>. Suppose that two friends on our
Campus want to meet in front of the Math building at noon-
time. Class locations and traffic being what they are, the
friends can only count on arriving sometime between noon
and 1 p.m. They agree to show up at the steps of the
building somewhere in that interval, with the stipulation
that, in order not to waste too much time waiting, each
will wait only 10 minutes after arriving and then leave
if the other has not shown up. What is the pr. of their
actually meeting?

To make this problem precise, we shall assume that
each of the two friends may arrive any time within the hour
from noon to 1 p.m., all times of arrival being distributed with the same uniform density (with L = 1 hour).
Thus, we have here life times (arrival times) of two
friends, to be denoted by X and Y, and each of these
random variables has the uniform distribution.

The second assumption to be imposed is that the arrival
times of the two friends are completely independent of each
other. This assumption that X and Y are independent
life times is of crucial importance; it is expressed by the

relation $f(x,y) = f_1(x)f_2(y)$. In the present case the joint density is

$$f(x,y) = 1/L^2, \qquad 0 \le x \le L, \qquad 0 \le y \le L$$

(and $f = 0$ outside the square with sides L).

Now, $X - Y$ is the distance between the friends arrivals. Since our problem requires that distance between arrivals be positive (and not exceed 10 min.) we must take the absolute value $|X-Y|$. The event A we look for is $|X-Y| \le h$, and its pr. is according to the above rule:

$$\Pr(|X-Y| \le h) = \iint_{|x-y| \le h} (1/L^2) \, dx \, dy$$

the region $R(x,y)$ in the plane being determined by inequalities:

$$|x-y| \le h, \qquad 0 \le x \le L, \qquad 0 \le y \le L.$$

Note that as a function of h, the above integral gives the d.f. of life time $|X-Y|$.

In order to evaluate this double integral, one could proceed formally, but in the present case a simple observation dispenses with any integration. First note that $1/L^2$ is a constant, so may be taken out from the integration sign; the integral then gives simply the area of the strip R. Next, notice that the strip R is obtained from the square of sides L, by removing two equal triangles of equal sides of length $L - h$. The area of a tringle being well known, we have:

$$\text{pr}(|X-Y| \le h) = (1/L^2)(L^2 - (L-h)^2) = 1 - (1-h/L)^2$$

for $0 \le h \le L$.

In our problem, $L = 1$ hour, $h = 10$ min. $= 1/6$ hour, so from the above formula, the required pr. is $11/36$.

Observe also that in the present case density is $(1-h/L)\ 2/L$, and the mean waiting time has been already calculated in Example 12 in Section 7, and equals $L/3 =$ 20 minutes.

2. <u>Quadratic equation</u>. We know from algebra that a quadratic equation

$$Ax^2 + Bx + Cx = 0$$

can have two, or one or none real roots. Suppose you write

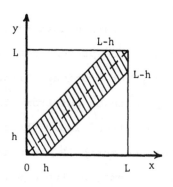

Fig 8.1

The event A that meeting takes place (shaded area).

down an equation choosing its coefficients at random. What
is the pr. that it has two real roots? Of course, you can
immediately check if the given equation has two real roots,
and no probability is needed. The problem is, however, more
general -- we actually ask how frequently you may expect an
equation to have two, one or no real roots. (The question
is far from being trivial, and for equations of higher
degree only approximate answers are known.)

We shall solve this problem in a very simple case.
First of all observe that coefficients A, B, and C are
random variables. Our first simplification will be to take
A = 1, a constant. Thus our equation can be rewritten as

$$x^2 + Bx + C = 0$$

where B and C are random variables. We need their joint
density. Here comes our second, rather restrictive assumption. Namely, we shall take B and C as independent,
each with uniform density over the unit interval [0,1].
Thus, the joint density is simply 1, over the unit square.

The quadratic equation has two real roots when
$B^2 - 4C > 0$; this is therefore the event whose pr. we must
evaluate. It is simply represented as the double integral
over the region R determined by the inequality

$$b^2 - 4c > 0, \qquad 0 \le b \le 1, \qquad 0 \le c \le 1$$

within the unit square. Hence:

$$\text{pr}(B^2 - 4C > 0) \;=\; \iint_{b^2 - 4c > 0} db \; dc \; .$$

The indicated region is bounded by the parabola $c = \frac{1}{4} b^2$ and the b-axis, and the above integral is simply the area of this region. Hence, its value is:

$$\int_0^1 \int_0^{\frac{1}{4}b^2} dc \;\; db \;=\; \frac{1}{4} \int_0^1 b^2 \; db$$

$$=\; \frac{1}{12} \; .$$

Hence the answer to our problem is 1/12.

Note that $B^2 - 4C = 0$ is the condition to have exactly one real root; the same argument gives the integral over the parabola $b^2 - 4c = 0$, and its value is of course 0. Consequently, it is unlikely to pick up an equation with only one root. On the other hand,

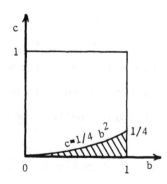

Fig 8.2
The region $b^2 - 4c > 0$ (shaded area).

the pr. of an equation having no real roots is obviously
11/12.

Clearly, this answer will change with a different
choice of distribution of coefficients.

3. <u>Double trouble</u>. When comparing two life times X and
Y it is of importance to know pr.'s of events like X > Y
or X = Y. These pr.'s are given by:

$$\text{pr}(X > Y) \quad = \quad \text{pr}(X-Y > 0) \quad = \quad \iint_{x-y>0} f(x,y) \; dx \; dy$$

$$\text{pr}(X=Y) \quad = \quad \iint_{x=y} f(x,y) \; dx \; dy \quad = \quad 0.$$

In the case of independence, the first integral reduces to

$$\int_0^\infty f_1(x) F_2(x) \; dx = 1 - \int_0^\infty f_2(y) F_1(y) \; dy. \quad \text{Suppose now that}$$

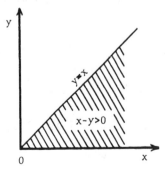

Fig 8.3

The event X > Y (shaded area).

X   and   Y   are independent with the same density   f.   Then
(with   z = F(x)):

$$Pr(X > Y) \quad = \quad \int_0^\infty f(x)F(x) \ dx$$

$$= \quad \int_0^1 z \ dz \quad = \quad \frac{1}{2} .$$

where   z = F(x), dz = f(x) dx, $0 \le x < \infty$, $0 \le z \le 1$.
It is of interest that this pr. is always $\frac{1}{2}$ for any density
f.

4.   <u>Proportion</u>.   Let X be time spent on shopping and let Y
be time spent on driving to a shopping mall.   Suppose one
wishes to compare time spent on shopping with total used
time, by considering the proportion

$$W = X/(X + Y).$$

Clearly, $0 \le W \le 1$.   Assume for simplicity that X and Y are
independent life times with the same exponential distribution (with mean $1/\lambda$.   Denote by G the d.f. of W.   Thus,

$$G(w) = pr(W \le w) = \iint \lambda e^{-\lambda x} \lambda e^{-\lambda y} \ dx \ dy =$$

$$= x/(x + y) \le w$$

$$= \int_0^\infty \lambda e^{-\lambda x} \ dx \int \lambda e^{-\lambda y} \ dy$$

$$(1-w)x/w$$

$$= \int_0^\infty \lambda e^{-\lambda x} \ dx \ e^{-\lambda(1-w)x/w} = w,$$

because the region R(x, y) is determined by inequality
x $\leq$ w(x + y) or (1 - w)x $\leq$ yw for x $\geq$ 0, y $\geq$ 0. Obviously,
0 $\leq$ w $\leq$ 1, and G(w) = 1 for w $\geq$ 1.

Thus, W has the uniform distribution over the interval
(0,1), with mean $\mathbb{E}$W = 1/2. which here coincides with $\mathbb{E}$X/
$\mathbb{E}$(X + Y), to confuse the unaware reader. However, it is
rather obvious that another proportion V = Y/(X + Y) also
has the same uniform distribution with mean 1/2. Indeed,
Problem 64 shows that when X and Y are independent with the
same distribution, then W and V must have their own common
distribution, too. Then, obviously, IEW = $\mathbb{E}$V = 1/2.

Section 9: How Normal Is Normal?

So far we have dealt exclusively with life times which
are nonnegative. Many situations require, however, introduction of negative values. For example, the difference of
two positive life times may assume negative values; also if
the time origin is within a life time, negative values
represent simply the time before the origin. In this Section a few comments on life times of arbitrary sign will be
made. It is more convenient to speak in such a situation
about a random variable (r.v.) rather than about life time;
see Section 2. If X is a r.v., its density f is now
defined for all x, from -∞ to +∞, and its d.f. F
now has this form:

$$F(x) = \int_{-\infty}^{x} f(t)\, dt, \quad -\infty < x < \infty$$

and the moments are defined by

$$\mu_n = \int_{-\infty}^{\infty} x^n f(x) \ dx.$$

Similarly, the average cost is now:

$$E \, \varphi(X) = \int_{-\infty}^{\infty} \varphi(x) f(x) \ dx.$$

This is actually where similarities end; other formulae which we discussed so far refer essentially to nonnegative life times.

$$* * * * * * * * * * * * * * *$$

We shall now restrict our discussion to the famous Normal (or Gaussian) distribution, of great importance in Statistics and Probability.

The normal density is defined for all real x by a single formula:

$$f(x) = \frac{1}{\sqrt{2\pi} \ \sigma} \ e^{-\frac{(x-\mu)^2}{2\sigma^2}}, \quad -\infty < x < \infty,$$

where μ and σ are two constants such that: $-\infty < \mu < \infty$ and $0 < \sigma < \infty$.

The graph of f is the familiar bell-shaped curve, symmetric with respect to the vertical line $x = \mu$. Indeed, $f(x) = f(2\mu - x)$. Furthermore, simple differentiation yields:

$$f'(x) = -\frac{x-\mu}{\sigma^2} f(x), \qquad f''(x) = \frac{1}{\sigma^4} [(x-\mu)^2 - \sigma^2] f(x) \ .$$

It follows that the curve $f(x)$ has maximum at $x = \mu$, equal to $f(\mu) = \dfrac{1}{\sqrt{2\pi}\ \sigma}$, and two inflection points at $\mu + \sigma$, and $\mu - \sigma$. Moreover, the width of the bell-shaped curve is proportional to the parameter σ; indeed, the distance between two inflection points is 2σ.

There is no nice formula for the d.f. F; one must write out the full integral. It is convenient to introduce the so called <u>standard</u> form of the normal density, denoted by ϕ, which is obtained by taking $\mu = 0$ and $\sigma = 1$:

$$\phi(x) = \frac{1}{\sqrt{2\pi}}\ e^{-\frac{x^2}{2}}, \qquad -\infty < x < \infty.$$

The corresponding standard form of the d.f. is of course

$$\Phi(x) = \int_{-\infty}^{x} \phi(t)\ dt .$$

Returning now to the arbitrary μ and σ, one finds that the d.f. F can be expressed in terms of a standard d.f. by the relation:

$$F(x) = \Phi((x - \mu)/\sigma) .$$

This follows by the change of variable $(t-\mu)/\sigma = y$.

The values of the standard distribution are tabulated so using tables one can compute $F(x)$ for any μ and σ. One can check that $F(\mu+\sigma) - F(\mu-\sigma) = 0.6826\ldots$.

****

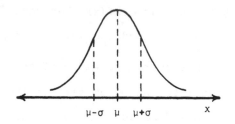

Fig 9.1

Gaussian density.

It is a rather complicated matter to verify that the integral of f over the whole region is unity, as it should. First of all one has

$$\int_{-\infty}^{\infty} f(x)\ dx\ =\ \int_{-\infty}^{\infty} \phi(t)\ dt\ =\ \frac{1}{\sqrt{2\pi}} \int_{-\infty}^{\infty} e^{-\frac{t^2}{2}}\ dt$$

$$=\ \sqrt{\frac{2}{\pi}} \cdot \int_{0}^{\infty} e^{-\frac{t^2}{2}}\ dt\ \equiv\ I.$$

Denote the last expression by I, and consider the product I^2 which can be represented by the double integral:

$$I^2\ =\ \frac{2}{\pi} \int_{0}^{\infty} e^{-\frac{y^2}{2}}\ dy \cdot \int_{0}^{\infty} e^{-\frac{z^2}{2}}\ dz\ =\ \frac{2}{\pi} \int_{0}^{\infty} \int_{0}^{\infty} e^{-\frac{z^2+y^2}{2}}\ dy\ dz.$$

To evaluate this integral, change variables $z = r \sin \alpha$, $y = r \cos \alpha$, $0 \le \alpha \le \frac{\pi}{2}$, $0 \le r < \infty$ so $dy\ dz = r\ dr\ d\alpha$, $y^2 + z^2 = r^2$, and:

$$I^2 = \frac{2}{\pi} \int_0^\infty \int_0^{\frac{\pi}{2}} e^{-\frac{r^2}{2}} r \, dr \, d\alpha = \int_0^\infty e^{-\frac{r^2}{2}} r \, dr$$

$$= \int_0^\infty e^{-s} \, ds = 1.$$

Since $I^2 = 1$, it follows that $I = 1$, because I is positive. This concludes the demonstration that f is a proper density.

$$* * * * * * * * * * * * * * * * *$$

It is much easier to identify the parameters μ and σ^2. Indeed, the following is true:

$$E(X) = \mu, \qquad var(X) = \sigma^2.$$

This follows from:

$$E(X) = \int_{-\infty}^\infty x f(x) \, dx = \int_{-\infty}^\infty (x-\mu) f(x) \, dx + \mu \int_{-\infty}^\infty f(x) \, dx$$

$$= \mu$$

by the symmetry of the normal density of f around μ. Next, by integration by parts:

$$var(X) = \int_{-\infty}^\infty (x-\mu)^2 f(x) \, dx = \sigma^2 \int_{-\infty}^\infty y^2 \phi(y) \, dy$$

$$= \sigma^2 \frac{1}{\sqrt{2\pi}} \int_{-\infty}^\infty y^2 e^{-\frac{y^2}{2}} \, dy = \frac{\sigma^2}{\sqrt{2\pi}} \int_{-\infty}^\infty y \, d(-e^{-\frac{y^2}{2}})$$

$$= \frac{\sigma^2}{\sqrt{2\pi}} \ [-ye^{-\frac{y^2}{2}} \Big|_{-\infty}^{\infty} + \int_{-\infty}^{\infty} e^{-\frac{y^2}{2}} \ dy \ ] \quad = \ \sigma^2 \int_{-\infty}^{\infty} \phi(y) \ dy$$

$$= \ \sigma^2 .$$

Thus, the gaussian distribution is completely characterized by its mean and variance.

****

A few more interesting integrals associated with the normal distribution. For simplicity suppose that $\mu = 0$. Then, by symmetry all moments of odd orders vanish, whereas moments of even orders are found to be:

$$\mu_{2k} \ = \ 1 \cdot 3 \cdot 5 \cdots (2k-1)\sigma^{2k}, \qquad \text{for} \quad k = 1,2,3,\ldots .$$

On the other hand, so called absolute moments of order r (where $r > 0$, not necessarily an integer) are (again for $\mu = 0$):

$$E(|X|^r) \ = \ \frac{2^{\frac{r}{2}}}{\sqrt{\pi}} \ \Gamma(\frac{r+1}{2})\sigma^r, \qquad \text{for any} \quad r > 0.$$

Obviously, both expressions agree for $r = 2k$.

****

Let X be a gaussian r.v. with mean μ and variance σ^2. Then, the pr. of the event $(a \le X \le b)$ can be computed from the following expression:

$$pr(a \leq X \leq b) \quad = \quad F(b) - F(a) \quad = \quad \Phi((b-\mu)/\sigma) - \Phi((a-\mu)/\sigma)$$

where F is the d.f. of X. This formula follows from the expression for F stated earlier.

Equivalently, this result can also be obtained by the analogous argument:

$$pr(a \leq X \leq b) \quad = \quad pr \left(\frac{a-\mu}{\sigma} \leq \frac{X-\mu}{\sigma} \leq \frac{b-\mu}{\sigma} \right) \quad = \quad pr \left(\frac{a-\mu}{\sigma} \leq Z \leq \frac{b-\mu}{\sigma} \right)$$

$$= \quad \Phi \left(\frac{b-\mu}{\sigma} \right) - \Phi \left(\frac{a-\mu}{\sigma} \right)$$

where $Z = (X-\mu)/\sigma$.

It can be shown that the r.v. Z has standard normal distribution, with mean 0 and variance 1.

Example 1: Let $\mu = 2$, $\sigma^2 = 9$ ($\sigma = 3$), $a = 2$, and $b = 5$. Then

$$pr(2 \leq X \leq 5) \quad = \quad \Phi \left(\frac{5-2}{3} \right) - \Phi \left(\frac{2-2}{3} \right) \quad = \quad \Phi(1) - \Phi(0)$$

$$= \quad \Phi(1) - \frac{1}{2} \quad = \quad 0.841 - 0.500 \quad = \quad 0.341$$

from table.

Remember: $\Phi(z)$ is the area under the standard normal curve $\phi(z)$ from $-\infty$ up to z.

Example 2: i) A r.v. X is normally distributed with mean zero and variance 1. Find a number such that

$$pr(|X| > a) \quad = \quad \frac{1}{2} .$$

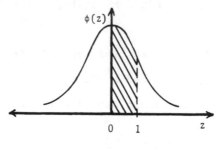

Fig 9.2

Using symmetry:

$$pr(|X| > a) \quad = \quad 1 - pr(|X| \le a) \quad = \quad 1 - 2pr(0 \le X \le a)$$

$$= \quad 1 - 2[\Phi(a) - \Phi(0)]$$

$$= \quad 1 - 2\Phi(a) + 2 \cdot \frac{1}{2} \quad = \quad 2 - 2\Phi(a) \quad = \quad \frac{1}{2}$$

so $\Phi(a) = \frac{3}{4}$. From the table: a = 0.675.

ii) A r.v. X is normally distributed with mean μ and variance σ^2. Find the expression for c such that

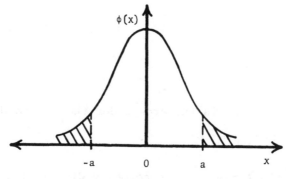

Fig 9.3

Example 2.

$$pr(|X-\mu| > c) \;=\; \frac{1}{2}$$

$$pr(|X-\mu| > c) \;=\; 1 - pr(|X-\mu| \le c) = 1 - pr(-c \le X-\mu \le c)$$

$$=\; 1 - pr\left(-\frac{c}{\sigma} \le \frac{X-\mu}{\sigma} \le \frac{c}{\sigma}\right)$$

$$=\; 1 - 2pr\left(0 \le \frac{X-\mu}{\sigma} \le \frac{c}{\sigma}\right) \;=\; 1 - 2\Phi\left(\frac{c}{\sigma}\right) + 1$$

$$=\; 2 - 2\Phi\left(\frac{c}{\sigma}\right) \;=\; \frac{1}{2}$$

so $\Phi\left(\frac{c}{\sigma}\right) = \frac{3}{4}$. Hence, $\frac{c}{\sigma} = a$ (from i), so $c = 0.675\sigma$.

⋆⋆⋆⋆⋆⋆⋆⋆⋆⋆⋆⋆⋆⋆⋆⋆

Remark: In applications, the gaussian approximation to pr.
of events (determined by other r.v.'s) turns out to be very
convenient for numerical computations. Its surprising
success in applications to problems in the physical sciences
justified the famous remark by Poincaré that there must be
something mysterious about the normal distribution because
"mathematicians think it is a law of nature whereas
physicists are convinced that it is a mathematical theorem."

For example, let X be a r.v. with mean $E(X) = \mu$ and
$var(X) = \sigma^2$. It is easy to check that the r.v.
$Z = (X-\mu)/\sigma$ has mean 0 and variance 1. The theorem
asserts that the distribution of Z can be approximated by
the standard gaussian distribution. Thus:

$$\text{pr}(a < X < b) \quad \approx \quad \Phi[(b-\mu)/\sigma] - \Phi[(a-\mu)/\sigma].$$

Exact conditions when such approximation is valid are expressed by the famous Central Limit Theorem, which unfortunately we cannot discuss here. It suffices to mention that as a first approximation, gaussian approximation is sufficiently accurate.

We shall now work out a rather not so convincing example, and postpone comments on the Central Limit Theorem to Subsection 10.3.

Example 3: For a r.v. X with mean μ and variance σ^2, compute

$$\text{pr}(\mu \leq X \leq 3\mu/2)$$

assuming that X is the exponential life time. Compare with the normal approximation.

$$\text{pr}(\mu \leq X \leq 3\mu/2) \;=\; e^{-\lambda\mu} - e^{-\frac{3}{2}\lambda\mu} \;=\; e^{-1} - e^{-\frac{3}{2}}$$

$$=\; 0.3679 - 0.2231 \;=\; 0.1448$$

because $\lambda\mu = 1$. Now:

$$pr(\mu \le X \le 3\mu/2) \quad = \quad pr(0 \le \frac{X-\mu}{\sigma} \le \frac{1}{2}\frac{\mu}{\sigma}) \quad = \quad pr(0 \le \frac{X-\mu}{\sigma} \le \frac{1}{2}),$$

because $\sigma^2 = \mu^2$ for n.e.d.

$$= \quad \Phi(\frac{1}{2}) - \Phi(0) \quad = \quad 0.6915 - 0.5000$$

$$= \quad 0.1915.$$

Section 10: When in Doubt, Approximate!

Frequently, a rough estimate of pr. of an event may be sufficient, instead of the exact value (especially, when exact calculations may be involved). The manner in which approximation is obtained depends of course on the formula to be used. However, in many situations one can obtain approximate formulae which are valid for any distributions or for a rather wide class of distributions.

10.1. One of the most useful formulae is the so called Chebyshev inequality which holds for any distribution having finite mean and (positive) variance. As an approximation, the Chebyshev inequality is rather crude, yet it gives a very handy tool both for practical and theoretical appli-cations. We shall discuss it in two versions, both of which are of interest.

Consider first a (non-negative) life time $Y \ge 0$, with (finite) mean $m = EY$. Let g be the density of Y. Then, for any positive number $a > 0$, we can write:

$$m = \int_0^\infty yg(y)\, dy \geq \int_a^\infty yg(y)\, dy \geq a \int_a^\infty g(y)\, dy$$

$$= a\ \mathrm{pr}(Y > a)$$

where the first inequality is obtained by reducing the range of integration, and the second follows from the fact that $y \geq a$ in this range; recall that $\mathrm{pr}(Y = a) = 0$.

Hence, for any life time Y and for any $a > 0$:

$$\mathrm{pr}(Y > a) \leq m/a.$$

This is the first version of our inequality. Clearly, it gives no information when the bound m/a is larger than 1.

Now let X be an arbitrary r.v. which may assume values of both signs, possessing a (finite) mean μ and (nonzero) variance σ^2. Clearly, a r.v. $Y = (X-\mu)^2$ is nonnegative and

$$EY = E(X-\mu)^2 = \sigma^2.$$

Hence, substituting into the above inequality for Y, one has:

$$\mathrm{pr}((X-\mu)^2 > a) \leq \sigma^2/a.$$

Writing $a = \varepsilon^2$, and noting that $(X-\mu)^2 > \varepsilon^2$ is equivalent to $|X-\mu| > \varepsilon$ we get the famous Chebyshev inequality.

Thus, for any r.v. X with finite mean μ and positive variance σ^2, one has for an arbitrary positive ε:

$$pr(|X-\mu| > \varepsilon) \leq \sigma^2/\varepsilon^2 \qquad (\varepsilon > 0).$$

Again, the approximation is useless when σ/ε is larger than 1. It is worthwhile to note that the Chebyshev inequality gives the bound in terms of variance, of the pr. of deviation from the mean being larger than $\varepsilon > 0$. The pr. in question is of course given by the integral

$$pr(|X-\mu| > \varepsilon) = \int_{-\infty}^{\mu-\varepsilon} f(x)\ dx + \int_{\mu+\varepsilon}^{\infty} f(x)\ dx$$

where f is the density of X. This integral gives the exact value, but the advantage of the Chebyshev inequality is that there is no need of evaluation of these integrals.

Example: i) Suppose that X has exponential life, with mean $1/\lambda$. Then, the exact value of the pr. of deviation, and its bound, are easily found to be:

$$pr(|X - 1/\lambda| > \varepsilon) = 1 - 2e^{-1} \sinh(\varepsilon\lambda) \leq (1/\lambda\varepsilon)^2$$

(for $\varepsilon > 1/\lambda$).

ii) For Gaussian r.v. X, it is easy to find that

$$pr(|X-\mu| > \varepsilon) = 1 - \Phi(\frac{\varepsilon}{\sigma}) + \Phi(\frac{-\varepsilon}{\sigma}) = 2[1 - \Phi(\frac{\varepsilon}{\sigma})] \leq \frac{\sigma^2}{\varepsilon^2}$$

for $\varepsilon > 0$.

****

<u>Note</u>: Letting $\varepsilon = k\sigma$, and taking the complement, the Chebyshev inequality may be written as:

$$pr\,(\,|X-\mu|\,\le \sigma k)\ \ge\ 1 - 1/k^2 \qquad \text{for}\ \ k > 0.$$

This means that the pr. is always larger than $1 - k^{-2}$ (i.e., close to 1) that X assumes values inside the interval from $\mu - k\sigma$ to $\mu + k\sigma$. This confirms that the variance is indeed a measure of the spread.

10.2. More important applications of the Chebychev inequality are illustrated by the following examples.

Suppose that we have several r.v.'s, say $X_1, \ldots, X_n$, and we are interested in the sum

$$S\ =\ X_1 + \ldots + X_n.$$

If individual X's are life times, S is the total life time of n items (cf. Section 5; important applications are discussed in later sections). Suppose for simplicity that the X's are independent and identically distributed (abbreviation i.i.d. is used for such a case). Then as indicated in Section 5:

$$ES = n\mu, \qquad \text{var}\ S = n\sigma^2$$

where μ and σ^2 are mean and variance, respectively, common to all X's.

The Chebyshev inequality gives the estimate for distribution of S:

$$pr(|S-n\mu| > \varepsilon) \leq n\sigma^2/\varepsilon^2.$$

Much more interesting is the consideration of the r.v. $\overline{X} = S/n$, that is

$$\overline{X} = (X_1 + \ldots + X_n)/n.$$

This may be regarded as some kind of average life time; in Statistics $\overline{X}$ is known as a sample mean. Hence

$$E\overline{X} = \mu, \qquad var\ \overline{X} = \sigma^2/n.$$

The Chebyshev estimate is:

$$pr(|\overline{X}-\mu| > \varepsilon) \leq \sigma^2/(n\varepsilon^2).$$

It is of great importance to observe that when $n \to \infty$, then the bound goes to zero, so

$$\lim_{n\to\infty} pr(|\overline{X}-\mu| > \varepsilon) = 0, \qquad \varepsilon > 0.$$

We express this fact by saying that $\overline{X}$ converges to μ in probability (as $n \to \infty$). This is the famous (weak) law of large numbers ("law of averages" in old fashioned terminology) for i.i.d. r.v.'s with finite variance.

****

10.3 A few comments on the normal approximation mentioned
in the previous Section 9 are now in order. Suppose that
$X_1, \ldots, X_n$ are i.i.d. r.v.'s with common mean μ and finite
variance σ^2. As in Section 5, the sum

$$S = X_1 + \cdots + X_n$$

has mean $n\mu$ and variance $n\sigma^2$. Hence, the r.v.,

$$Z_n = \frac{S - n\mu}{\sigma\sqrt{n}},$$

has $\mathbb{E}X_n = 0$ and var $Z_n = 1$. The Central Limit Theorem asserts that, as $n \to \infty$, the distribution of Z_n tends to the
standard normal distribution (irrespective of the form of
common distribution of X's):

$$\lim_{n \to \infty} pr(Z_n \leq z) = \Phi(z), \qquad -\infty < z < \infty.$$

From this important result, one may obtain approximations for
distribution of the sum S and of the sample mean $\overline{X}$. Indeed,
one has

$$pr(a \leq S \leq b) \approx \Phi(z_2) - \Phi(z_1)$$

where

$$z_2 = (b - n\mu)/\sigma\sqrt{n}, \qquad z_1 = (a - n\mu)/\sigma\sqrt{n},$$

and

$$pr(a \leq \overline{X} \leq b) \approx \Phi(z_2) - \Phi(z_1)$$

where

$$z_2 = (b - \mu)\sqrt{n}/\sigma, \qquad z_1 = (a - \mu)\sqrt{n}/\sigma.$$

Indeed, in the first case,

$$pr(a \leq S \leq b) = pr(z_1 \leq Z_n \leq z_2) \approx \Phi(z_2) - \Phi(z_1)$$

and in the second case,

$$pr(a \leq \overline{X} \leq b) = pr(z_1 \leq Z_n \leq z_2) \approx \Phi(z_2) - \Phi(z_1).$$

Of great practical interest is the accuracy and speed of these approximations. This is a fascinating subject, but we cannot discuss it here and must satisfy ourselves with the meager example from the previous section.

Chapter 1: Problems

1. Suppose that a life time X has density f given by:

$$f(x) = \begin{cases} \dfrac{2}{a^2}(a-x) & \text{for} \quad 0 \leq x \leq a \\[2mm] 0 & \text{for} \quad x \geq a \end{cases}$$

where a is a positive constant.

(a) Show that the d.f. F is given by

$$\cdot F(x) = \begin{cases} (\dfrac{x}{a})(2 - \dfrac{x}{a}) & \text{for} \quad 0 \leq x \leq a \\[2mm] 1 & \text{for} \quad x \geq a \end{cases}$$

(b) Show that the mean life time μ is

$$\mu = \frac{a}{3}$$

and that its variance is

$$\sigma^2 = \frac{a^2}{18} .$$

(c)   Sketch graph of $f(x)$, $F(x)$,   indicating slopes, etc.

(d)   Take $a = 2$ hours, and compute the pr. that the life time is between 30 and 60 minutes.

2.   Suppose that reading habits of a person are described by the following density of a time X needed to read a book (with the maximum time allotted to be L hours):

$$f(t) = c \frac{t}{L} , \quad \text{for } 0 \le t \le L$$

where c is a constant to be determined. Show that:

(a) $c = \frac{2}{L} .$

(b)   The pr. of time needed being t or less is

$$F(t) = (\frac{t}{L})^2 , \quad \text{for } 0 \le t \le L.$$

(c)   The pr. that the time needed is between a and b is $\frac{b^2 - a^2}{L^2} .$

(d)   Moments of order n are $\mu_n = \frac{2}{n+2} L^n$ $(n = 0,1,2,...)$.

(e)   Mean reading time is $\frac{2}{3} L$, and the variance is $\sigma^2 = \frac{L^2}{18} .$

(f)   The pr. that the reading will continue an addition-

al h hours, given that it already lasted more
than t hours is:

$$\frac{L^2 - (t+h)^2}{L^2 - t^2} , \qquad \text{for} \quad 0 \le h \le L-t.$$

(g) Find the probability that the time needed to read
the book will be less than average but more than
half of the allotted time.

(h) Find the probability that the time needed to read
the book will be more than average, when it is
known that half of the allotted time has already
elapsed.

(i) Suppose that reading of the book has been inter-
rupted at some random instant. Show that the
average time which would be needed to complete
the reading (from that random instant) is (3/8)L.

3. Suppose that the life time X of a TV tube is described
by the following density f, with the maximum life time
equal to L years:

$$f(x) = \begin{cases} cx(L-x), & \text{for} \quad 0 < x < L \\ 0 & \text{otherwise} \end{cases}$$

where c is a constant to be determined. Show that

(a) $c = 6/L^3$.

(b) The d.f. F is given by

$$F(x) = \begin{cases} [3-2(x/L)](x/L)^2, & \text{for} \quad 0 \le x \le L \\ 1, & \text{for} \quad L \le x < \infty \end{cases}$$

and F has the inflection point at $x = \frac{1}{2}L$,

with $F(\frac{1}{2}L) = \frac{1}{2}$.

(c)   The pr. that the tube will last more than   a,

but less than   b,   is given by:

$$pr(a < X < b) = \frac{b-a}{L^3} [a(3L-2a-b) + b(3L-2b-a)]$$

for $0 \le a \le b \le L$.

(d)   Moments of order   n   are:

$$\mu_n = 6L^n/(n+2)(n+3), \quad \text{for}\ \ n = 0,1,2,\dots .$$

(e)   Mean life time is $\frac{1}{2}L$, $var(X) = L^2/20$.

(f)   The pr. that the tube will last an additional   h

years, given it already lasted more than   x   years,

is:

$$\frac{F^C(x+h)}{F^C(x)} = \frac{1 - 3(\frac{x+h}{L})^2 + 2(\frac{x+h}{L})^3}{1 - 3(\frac{x}{L})^2 + 2(\frac{x}{L})^3}$$

for $h \le L-x$.

(g)   Suppose that the mean life time of a tube is   5

years.

(i)   Plot graphs of   f   and   F.

(ii)   Compute the standard deviation of the life
time.

(iii)   Compute the pr. in (c) for   a = 2   years and
b = 3   years.

(iv)   Compute the pr. in (f) for   x = 2   years and
h = 1   year.

4. The scheduled length of a meeting is b (minutes, say). It is unlikely that the meeting will end before a (minutes), and it is impossible that it will last more than L (another meeting is scheduled in the same room). Here $0 < a < b < L$.

Assume that the density f of the life time X (duration of the meeting), corresponding to the above restrictions, has the form:

$$f(x) = \begin{cases} 0 & \text{for } 0 < x < a \\[2mm] 2\beta\dfrac{x-a}{(b-a)^2} & \text{for } a < x < b \\[2mm] \dfrac{1-\beta}{L-b} & \text{for } b < x < L \\[2mm] 0 & \text{for } L < x < \infty \end{cases}$$

Here β is a constant such that $0 < \beta < 1$.

(a) Show that the d.f. F is given by:

$$F(x) = \begin{cases} 0 & \text{for } 0 \le x \le a \\[2mm] \beta\left(\dfrac{x-a}{b-a}\right)^2 & \text{for } a \le x \le b \\[2mm] \dfrac{1-\beta}{L-b}\,x + \dfrac{\beta L-b}{L-b} & \text{for } b \le x \le L \\[2mm] 1 & \text{for } L \le x < \infty \end{cases}$$

(b) Show that the average length of the meeting E(X) is given by:

$$\mu = \frac{\beta}{3}(2b+a) + \frac{1-\beta}{2}(L+b).$$

(c)   For given   a, b   and   L,   find   β   for which
      μ = b.   Verify that indeed   0 < β < 1,   and that
      always   3L − b − 2a > 0.   What when   β = 1 ?

(d)   Take: a = 45 min.,   b = 50 min.,   L = 60 min.,
      and   β = 9/10.   Plot graphs of   f(x)   and   F(x);
      indicate slope, discontinuities (if any), etc.
      Compute   E(X).

(e)   Find   β   for which   E(X) = b,   as in (c) above,
      using   a, b   and   L   for (d) above.

(f)   Using the values from (d) above compute:
      The pr. that the meeting will terminate before
      50 minutes,

      The pr. that the meeting will terminate after
      50 minutes,

      The pr. that the meeting will terminate within
      the interval   48-51   minutes.

5.   Suppose that the hazard rate of a life time   X   is
     given by:

$$r(t) \;=\; \frac{a}{b+t}\,, \qquad 0 \le t < \infty$$

     where   a > 1   and   b > 0   are constants.
     Show that:

(a)   The corresponding d.f. of   X   is given by:

$$F(x) \;=\; 1 - \left(\frac{b}{b+x}\right)^{a}, \qquad 0 \le x < \infty.$$

(b) The mean life time and variance are:

$$\mu = \frac{b}{a-1}, \qquad \sigma^2 = \frac{a}{a-2} \mu^2 \qquad \text{(for } a > 2)$$

so $\sigma^2/\mu^2 = a(a-2)^{-1} > 1.$

(c) The c.d.f. of prolongation by h is:

$$K_t(h) = 1 - \left(\frac{b+t}{b+t+h}\right)^a, \qquad 0 \le h < \infty$$

with density

$$k_t(h) = \frac{a}{b+t} \cdot \left(\frac{b+t}{b+t+h}\right)^{a+1}.$$

(d) Take $\mu = 20$ minutes, and $a = 4$. Plot graphs of $r(t)$ and $F(x)$.

6. Suppose that in some economics problem the hazard rate r is zero as long as life (say, income) stays below a threshold value b. At b, the rate jumps to a fixed value, and then decreases inversely proportionally to the life.

Thus, the hazard rate has the form:

$$r(x) = \begin{cases} 0 & \text{for } 0 \le x < b \\ \beta/x & \text{for } b \le x < \infty, \end{cases} \qquad \begin{array}{l} \text{where } \beta \text{ is} \\ \text{a constant,} \\ \beta > 1 \end{array}$$

(a) Compute the hazard function R for all $x \ge 0$.

(b) Show that the corresponding life time d.f. F is given by:

$$F(x) = \begin{cases} 0 & \text{for } 0 \le x \le b \\ 1 - (b/x)^{\beta} & \text{for } b \le x < \infty \end{cases}$$

(Note: this is known in Economics as Pareto distribution of income).

(c) Show that the average life time is

$$\mu = \frac{b\beta}{(\beta-1)}, \quad \text{and} \quad \sigma^2 = \frac{\mu^2}{\beta(\beta-2)} \quad \text{(for } \beta > 2).$$

7. (a) Assume that the life time of a battery in a flashlight has the uniform distribution with constant density $1/L$. Suppose that the mean was found to be 25 days. Determine L.

(b) Suppose that the battery has already lasted 30 days. Find the probability that the flashlight will continue to operate (with the same battery) for more than one week.

8. Suppose that a clerk made a record of time spent waiting for a bus during a week when going to work in the morning, and found that the mean waiting time was 10 minutes. Find the average time between buses and determine the probability of waiting between 5 and 19 minutes.

Consider two cases according to the assumption that the time interval between buses is distributed uniformly or exponentially.

9. In the formula for the mean waiting time w in the bus
    problem, assume that the life time distribution has the
    property that the standard deviation is proportional
    to the mean, that is $\sigma = k\mu$, where k is some
    constant.

    (a) Express w as a function of k and of μ.

    (b) Assume μ fixed, and plot the graph of w as a
         function of k. Indicate for which values of k
         the following inequality holds, $w > \mu$.

10. Suppose that inter-arrival times between buses have
    density:

    $$f(t) \;=\; \begin{cases} 2t/L^2, & 0 \le t \le L \\ 0, & L < t \end{cases}$$

    Show that the waiting time for the bus has d.f.:

    $$G(t) \;=\; \begin{cases} \dfrac{3}{2}\dfrac{t}{L}\,[1 - \dfrac{1}{3}(\dfrac{t}{L})^2], & 0 \le t \le L \\ 1, & L \le t \end{cases}$$

    and the mean waiting time is $w = \dfrac{3}{8} L$.

11. Suppose that the amount of time one spends in a bank
    is distributed with mean 10 minutes.

    (a) What is the pr. that a customer <u>arriving at
         random</u>, will spend more than 10 minutes in
         the bank?

(b)   What is the average time in the bank, for the
      customer in (a) ?

Consider three cases separately, according to
distribution of the amount of time being

(1)   uniform

(2)   exponential

(3)   same as in Problem 5.

Give numerical results.

12.   Suppose that the amount of time that a light bulb
      works before burning itself out is uniformly distrib-
      uted with mean μ.   Suppose that a person enters a
      room in which a light-bulb is burning for some time.
      What is the pr. that he/she will be able to use that
      light for more than the average μ ?

13.   Suppose that independent life times X and Y have
      exponential distributions, but with <u>different</u> parameters,
      λ and μ respectively.

      (a)   Show that the life time $Z = X + Y$ has density
            g of the form:

            $$g(z) \;=\; \frac{\lambda\mu}{\mu-\lambda}\,(e^{-\lambda z} - e^{-\mu z}) \qquad \text{for } \lambda \neq \mu.$$

      (b)   Verify that the d.f. G of Z is

            $$G(z) \;=\; 1 - \frac{\mu}{\mu-\lambda}\,e^{-\lambda z} + \frac{\lambda}{\mu-\lambda}\,e^{-\mu z} \qquad \text{for } \lambda \neq \mu.$$

(c) Verify that $EZ = EX + EY = 1/\lambda + 1/\mu$.

(d) Deduce that for $\lambda = \mu$, the above expressions reduce to that in Example 1, Section 5.

14. Suppose that the joint density of two life times X and Y has the form:

$$f(x,y) = \begin{cases} c(x+y) & \text{for } x, y \text{ in region } T \\ 0 & \text{for } x, y \text{ outside } T \end{cases}$$

where T is a triangle in the first quadrant determined by lines:

$$x = a, \qquad y = 0, \qquad y = mx$$

where a and m are positive constants.

(a) Show that $c = 6a^{-3}(2m+m^2)^{-1}$.

(b) Evaluate $pr(\frac{1}{2}a \le X \le a, \ 0 \le Y \le \frac{1}{2}am)$.

(c) Find marginal densities $f_1(x)$ and $f_2(y)$, and verify that X and Y are dependent.

15. In this problem the life time X has density f of the form given in Problem 3. Suppose that the gain from operating a certain equipment depends on the duration of its life time in the following way. In the initial stages of operation when life does not exceed a fixed length a, gain increases proportion-

ally to the length of life. Then when operation is established, gain remains constant throughout the remaining period from a to L.

More precisely, the cost function has the following form:

$$\varphi(x) = \begin{cases} mx & \text{for } 0 \le x \le a \\ ma & \text{for } a \le x \le L \end{cases} \quad \text{where} \quad \begin{array}{l} 0 \le a \le L \\ 0 \le m < \infty \end{array} \quad \begin{array}{l} \text{are} \\ \text{constant} \end{array}$$

(a) Show that the average gain is:

$$E\varphi(X) = \frac{1}{2} ma[2 - 2(a/L)^2 + (a/L)^3].$$

(b) Suppose now that the slope m is related to the position a in a manner described by the following relation:

$$ma = c(5a+4L) \qquad \text{for } 0 \le a \le L,$$

where c is a fixed constant (measured in dollars per unit of life).

Regard the average gain $E\varphi(X)$ as a function of the location a, and show that gain is maximum for the following choice of a:
$a_m = \frac{1}{2}L$, and the value of this maximum is $(169/32)cL$.

Hint: Use the following factorization of the cubic:

$$20z^3 - 18z^2 - 16z + 10 = 2(5z^2-2z-5)(2z-1)$$

for $0 \le z \le 1$.

(c)   Select L = 10 years, and plot the following graphs
      as functions of a: (take c = 2 hundred dollars
      per year) Hint: for convenience use: z = a/L.

      (i)   ma; for selected values of a, indicate
            the corresponding cost function.

      (ii) m; what are values of m at a = 0,
            $a = \frac{1}{2}L$,   a = L ?

      (iii) Average gain Eφ(X), indicating the maximum.

16.   Suppose that the life time X has density f of the
      form given in Problem 3. Suppose that when the life
      time is smaller than some value t, where 0 < t ≤ L,
      then the cost of operation is constant but inversely
      proportional to t.

      If the life time is larger than t, the cost is
      zero. Thus, the cost function is

$$\varphi(x) = \begin{cases} \dfrac{cL}{t} & \text{for } 0 \le x \le t \\ 0 & \text{for } t < x \le L \end{cases}$$

      where c is a positive scaling constant.

      (a) Show that the average cost is:

           $$\mathbb{E}\,\varphi(X) = c\left(\frac{t}{L}\right)\left(3 - 2\frac{t}{L}\right) .$$

      (b) Let now t be a variable. Plot the graph of the
           average cost Eφ(X) as a function of t, and
           show that it has a maximum at t = (3/4)L, equal
           to (9/8)c.

17. Maintaining cost of a certain equipment depends on the
    duration of its life in the following way. In the
    initial stages of operation when life does not exceed
    a fixed length a, cost increases sharply, proportion-
    ally to the square of life. Then when operation is
    established cost remains constant throughout the whole
    period from a to some other point b. When equipment
    is old, that is when its life is over b, cost increases
    rather slowly with constant rate.

    More precisely, the cost function has the follow-
    ing form:

$$\varphi(x) = \begin{cases} \lambda^2 x^2 & \text{for } 0 \le x \le a \\ \lambda^2 a^2 & \text{for } a \le x \le b \\ m(x-b) + \lambda^2 a^2 & \text{for } b \le x < \infty \end{cases}$$

$$a < b, \quad m > 0, \quad \lambda > 0.$$

Furthermore, it is assumed that life time X is
exponential, with mean $\mu = 1/\lambda$.

(a) Draw the graph of the cost function.

(b) Show that the average cost is

$$E\varphi(X) = 2 - 2(\lambda a + 1)e^{-\lambda a} + \frac{m}{\lambda} e^{-\lambda b}.$$

(c) Suppose now that the period of the constant cost
     operation has fixed duration h, so b = a + h.

Regard the average cost $E\varphi(X)$ as a function of the location a, and show that the cost is minimum for the following choice of a:

$$a = \frac{m}{2\lambda^2} e^{-\lambda h}$$

and the value of this minimum cost is $2(1-e^{-\lambda a})$, independent of length h.

(d) Select the average life $\mu = 10$ days, $h = 20$ days, and take $m = \frac{1}{2}$. Plot the graph of the cost $E\varphi(X)$ as the function of a, indicating the position of the minimum.

18. Suppose that the life time X has the density f given in Problem 1, with mean $\mu = \frac{a}{3}$.

Suppose that the cost of operation of the system is such that whenever the life stays below its mean value, the cost is proportional to the length of life, plus a fixed initial cost.

Whenever the life overshoots its mean value, the cost remains constant and equal to its value when life is equal to its mean.

In symbols, the corresponding cost function φ is given by:

$$\varphi(x) = cx + b \quad \text{for} \quad 0 \le x \le \mu,$$
$$\varphi(x) = c\mu + b \quad \text{for} \quad \mu \le x \le a,$$

where $c \geq 0$ is the cost per unit of time and
$b \geq 0$ is a fixed initial cost.

(a)   Show that the average cost $E\varphi(X)$ is given by:

$$E\varphi(X) \quad = \quad \frac{19}{27} c\mu + b.$$

(b)   Take $\mu = 9$ hours, $c = 6$ dollars per hour,
      $b = 14$ dollars. Compute $E\varphi(X)$.

(c)   (i)   For fixed c and b, plot the graph of the
            average cost $E\varphi(X)$ as a function of the
            mean μ.

      (ii) Plot the graph of the cost function $\varphi(x)$,
            as a function of x.

19.   An old lady noticed that when her favorite tea pot is
      younger than the average life of tea pots, it produces
      good tea. If, however, its life is larger than the
      average, then the taste of tea deteriorates proportion-
      ally to the age of the tea pot. Express the lady's
      discomfort by the "cost function" $\varphi(x)$ which equals
      0 when $x < \mu$, and equals ax when $x \geq \mu$, where
      a is constant, and μ is the mean.

          Assuming that tea pot life is exponential, show
      that the lady's average discomfort $E\varphi(X)$ equals
      $2a\mu e^{-1}$.

20. Consider a situation when an animal contacts a dis-
     ease which incapacitates it for a fixed length h > 0
     of time; afterwards the animal becomes immune. Sup-
     pose that the animal's incapacity stays constant during
     the illness (measured in some convenient units), whose
     magnitude increases with the animal's age   a   at the
     time of infection.

          Thus the "illness function" is given by

$$\varphi(x) = ca \qquad \text{for} \quad a < x < a+h$$

     and $\varphi(x) = 0$ otherwise. Here c is a positive
     constant.

          Assume (for simplicity) that the animal's life
     is exponential.

          Compute the "average illness," and show that it
     reaches a maximum when the infection instant equals
     the average life time. Sketch a graph of the "average
     illness" as a function of a. What would be the
     meaning of $h \to \infty$?

21. Suppose that the density f of a life time X is
     <u>symmetric</u> around a point c > 0, that is: $f(c-t) =$
     $f(c+t)$ for all t such that $0 < t < h$ where h
     is a constant such that $c-h > 0$, and f = 0 outside
     the interval (c-h, c+h).

          Show that the mean EX equals c.

22.   Suppose that the cost function φ has the form:

$$\varphi(x) = \begin{cases} a & \text{when } x \leq t \\ b & \text{when } x > t \end{cases}.$$

Show that the average cost is $aF(t) + bF^c(t)$. Plot
its graph as a function of t. What if a = b ?
Is there any maximum or minimum?

23.   Let f and F be the density and d.f., respectively,
of a life time X. Consider the cost function defined
by: $\varphi(x) = F(x)$. Verify that $\mathbb{E}\varphi(X) = \frac{1}{2}$ always!

24.   Let R be a hazard function, and let X be a life time
with distribution determined by R. Consider the cost
function defined by $\phi(x) = R(x)$. i) Verify that
$\mathbb{E}\phi(X) = 1$ always. ii) Assuming that R is strictly increasing, show that R(X) has exponential distribution.

25.   Suppose that the hazard rate r has the form:

$$r(t) = \begin{cases} c \cdot \sin(\frac{\pi}{L}t) & \text{for } 0 \leq t \leq L \\ c & \text{for } L < t < \infty \end{cases}$$

where L and c are positive constants.
      Show that the d.f. F of the life time has the
form

$$F(x) = 1 - e^{-R(x)}$$

where

$$R(x) = \begin{cases} \dfrac{cL}{\pi} (1 - \cos \dfrac{\pi x}{L}) & \text{for } 0 \le x \le L \\[3mm] \dfrac{2cL}{\pi} + c(x-L) & \text{for } L \le x < \infty. \end{cases}$$

Let the cost function $\varphi = 0$ for $t < L$, and $\varphi = a$ for $t > L$ where $a = \dfrac{2cL}{\pi}$. Show that the average cost is ae^{-a}.

26. A manufacturer produces two models of an apparatus, say Model A and Model B. Let X be the life time of Model A, and Y the life time of Model B. For simplicity, assume that X and Y are independent exponential lifes with parameters λ and μ, respectively.

    (a) In order to compare these models, the manufacturer is interested in the ratio Y/X of their life times. Show that for $z \ge 0$:

    $$\mathrm{pr}(Y/X \le z) = \mu z/(\mu z + \lambda).$$

    What is the average ratio $E(Y/X)$? Is there anything strange about it? Explain.

    (b) A comparison of models may be based on the pr. that Model A lasts longer than Model B. Show that

    $$\mathrm{pr}(Y \le X) = \mu/(\mu+\lambda).$$

(c) Suppose that from the records of sales it was
     estimated that the average life of Model B is 3
     times larger than the average life of Model A.
     Compute the above probability in this case.

(d) Comment on agreement or discripancy between results
     obtained in (c).

27. Let X be the life time of a new device, and let Y
     be the life time of an old device. Assume X and Y
     are independent and identically distributed with the
     common d.f. F with density f. In order to compare
     these devices, it is required to evaluate the proba-
     bility that $Y > X$, but with the restriction that
     $Y \leq a$, where a is fixed.

     (a) Write down the joint density $f(x,y)$ of X and
          Y.

     (b) Indicate the region in the plane corresponding
          to the event $X < Y \leq a$.

     (c) Set up the double integral for the required
          probability.

     (d) Evaluating this integral, show that its value is
          $(1/2)[F(a)]^2$, for arbitrary F.

28. Let X and Y be two independent life times, each
     having exponential distribution with parameter λ and

μ, respectively. Assume $\lambda \neq \mu$ (for $\lambda = \mu$ see Section 6).

(a) Show that the life time $Z = \min(X,Y)$ has also exponential distribution but with parameter $(\lambda + \mu)$.

$$\text{Deduce that } \quad E(Z) \leq \frac{1}{2}[E(X) + E(Y)]$$

(b) Show that the life time $T = \max(X,Y)$ has d.f. G of the form:

$$G(t) = 1 - e^{-\lambda t} - e^{-\mu t} + e^{-(\lambda+\mu)t}, \qquad t > 0$$

with density

$$g(t) = \lambda e^{-\lambda t} + \mu e^{-\mu t} - (\lambda+\mu)e^{-(\lambda+\mu)t}, \quad t > 0$$

and moments

$$ET^n = n! \left[\frac{1}{\lambda^n} + \frac{1}{\mu^n} - \frac{1}{(\lambda+\mu)^n} \right], \qquad n = 0,1,\dots \;.$$

29. A car owner installed two new tires of different brands. Let X be the life time of the first tire assumed to have uniform distribution over $(0,a)$. Let Y be the life time of the second tire assumed to have uniform distribution over $(0,b)$. Assume that X and Y are independent. Suppose also that $b < a$.

(a) Write down the joint density $f(x,y)$ of X and Y.

(b)   Plot the rectangle in the plane in which X   and
      Y   take their values jointly, and indicate the
      region corresponding to the event (Y > X).

(c)   Show that the probability of Y being larger than
      X   is b/(2a).

(d)   Compute this probability when it is known that the
      mean of X is twice the mean of Y.

30.   Let X be the life time of an engine in a car, and let
      Y be the life time of the body of a car. Assume (for
      simplicity) that X and Y are independent life times
      having n.e.d. with parameters λ and μ, respectively.
          As long as the engine is OK (i.e., when $X \geq Y$),
      one can assume that eventual damage to the body of the
      car is expressed by a fixed number c (in dollars, say).
      On the other hand, when X < Y the utility of the car
      drops down to zero. In other words, the utility func-
      tion φ is of the form:

$$\varphi(x,y) \;=\; \begin{cases} c & \text{when } x \geq y \\[2mm] 0 & \text{when } x < y \end{cases} \qquad (x \geq 0, \; y \geq 0, \; c > 0)$$

Show that the average utility is:

$$E\varphi(X,Y) \;=\; c \iint_{x \geq y} \lambda e^{-\lambda x} \mu e^{-\mu y} \, dx \, dy \;=\; c\mu/(\lambda+\mu).$$

31. A person must go to a drug store and to a bookshop; let X and Y be time spent in these two places, respectively. Assume that X and Y are independent r.v.'s having the same n.e.d. with a parameter λ.

   Suppose that this person can do this shopping only if the total time spent, X + Y, is t or less (where t is fixed); imagine then that the person receives a reward c under such conditions. When X + Y is larger than t, the reward is zero (due to missing the next appointment, say). That is, the reward function φ is of the form:

$$\varphi(x,y) \;=\; \begin{cases} c & \text{when } x+y \le t \quad\;\; (c > 0) \\ 0 & \text{when } x+y > t \end{cases}$$

Show that the average reward is:

$$E\,\varphi(X,Y) \;=\; \lambda^2 c \iint_{x+y\le t} e^{-\lambda(x+y)}\; dx\; dy$$

$$=\; c[1 - (1+\lambda t)e^{-\lambda t}]$$

and verify that this is an increasing function of t.

32. The error in a certain angle-measuring device has been found to be normally distributed with mean one minute and standard deviation σ of three minutes.

   (a) What is the pr. that a measurement is in error by more than three minutes?

(b)   Repeat if one minute is subtracted from the measured value.

33.   The lifetimes of two competing brands of vacuum tubes
can be viewed as normally distributed r.v.'s. Brand A
has mean life of 27 hours and a standard deviation of
5 hours, whereas brand B has $\mu = 30$ and $\sigma = 2$ hours.

(a)   Which brand should be chosen for use in an experimental aircraft with a flight time of 30 hours?

(b)   Same question for 34 hours.

(c)   Show that the pr. of a negative lifetime for each
brand is negligible (this can be interpreted as
the pr. that a tube is no good on delivery).

34.   Assume that the life in hours of a flashlight is normally distributed with a mean of 100 hours. If a purchaser requires at least 90% of them to have lives
exceeding 80 hours, what is the largest value that
the standard deviation σ can have and still have the
purchaser satisfied?

35.   (a)   Show that for the standard normal distribution:

$$\Phi(-x) = 1 - \Phi(x), \quad \text{for any } x \geq 0.$$

Show this relation on the graph of density ϕ.

(b)   Let X be a Gaussian r.v. with mean μ and
standard deviation σ. Show that

$$pr(|X| > a) = 2 - \Phi((a+\mu)/\sigma) - \Phi((a-\mu)/\sigma)$$

for any positive a. Indicate this relation on
the graph of density of X.

36.   Suppose that measurements in a certain experiment can
be represented by a Gaussian random variable of the
form $X = c + E$, where c is a constant value and E
is the Gaussian error with mean 0 and variance σ^2.

(a)   Show that the mean and the variance of X are c
and σ^2, respectively.

(b)   Show that the pr. of good performance, i.e. the pr.
that the (absolute) error does not exceed some
$\varepsilon > 0$, is:

$$pr(|E| \leq \varepsilon) = 2\Phi(\varepsilon/\sigma) - 1.$$

(c)   Verify that $\varepsilon < \sigma$ implies that this pr. is less
than 0.682.

37.   In a certain equipment there is a critical value c of
the life time X of the equipment, such that when
fluctuation of X around c does not exceed (in absolute value) c itself, then equipment performs well.
(Here $c > 0$ is a constant).

Suppose that X has d.f. F, and mean μ and variance σ^2.

(a) Compute the pr. of good performance, i.e. show that

$$pr(\,|X-c|\,\le c) \;=\; F(2c).$$

(b) Show that the Gaussian approximation to this pr. is

$$\Phi(\frac{2c-\mu}{\sigma}) \;+\; \Phi(\frac{\mu}{\sigma}) \;-\; 1.$$

(c) Assuming that $\mu = 2c$ and $\sigma^2 = c^2$, show that the pr. in (b) equals 0.477 ... .

38. Let X be a Gaussian r.v. with mean μ and variance σ^2. Consider the following life time:

$$Y \;=\; |X - \mu|$$

(absolute fluctuation around the mean). Show that Y has "folded Gaussian" density given by

$$f(y) \;=\; \sqrt{\frac{2}{\pi}}\,\frac{1}{\sigma}\,e^{-\frac{y^2}{2\sigma^2}}, \qquad\qquad \text{for } y > 0$$

and $f = 0$ for $y < 0$.

Verify that:

$$E(Y) \;=\; \sqrt{\frac{2}{\pi}}\,\sigma, \qquad var(Y) \;=\; \sigma^2(1 - \frac{2}{\pi}).$$

39. Let X be a Gaussian r.v. with mean μ and variance σ^2. Consider the following life time:

$$U = (X-\mu)^2$$

(square variation around the mean): Show that U has "chi-square" density given by

$$f(u) = \frac{1}{\sqrt{2\pi u}\sigma} e^{-\frac{u}{2\sigma^2}}, \qquad \text{for } u > 0$$

and $f = 0$ for $u < 0$.

Verify that:

$$E(U) = \sigma^2, \qquad \text{var}(U) = 2\sigma^4, \qquad E(\sqrt{U}) = \sqrt{\frac{2}{\pi}}\,\sigma.$$

40. Let X and Y be two independent Gaussian r.v.'s with mean μ_1, μ_2 and variance σ_1^2, σ_2^2, respectively. Show that a r.v. $Z = X + Y$ is also Gaussian with mean $\mu_1 + \mu_2$ and variance $\sigma_1^2 + \sigma_2^2$.

(Hint: see Sections 5 and 9; complete the square in convolution integral.)

41. Suppose that the life time X has a d.f. F of the form:

$$F(t) = 1 - \sum_{k=1}^{n} a_k e^{-\lambda_k t}, \qquad t \geq 0$$

where $a_k > 0$, $\lambda_k > 0$ and

$$\sum_{k=1}^{n} a_k = 1.$$

(i)   Verify that $F(t)$ is a d.f., with $F(0) = 0$,
$F(\infty) = 1$,   with density

$$f(t) \;=\; \sum_{k=1}^{n} a_k \lambda_k e^{-\lambda_k t}, \qquad t > 0.$$

(ii)   Show that the average life time is:

$$E(X) \;=\; \sum_{k=1}^{n} a_k / \lambda_k .$$

(iii)   Show that the Laplace transform is:

$$f^*(\alpha) \;=\; \sum_{k=1}^{n} \lambda_k a_k \, / \, (\lambda_k + \alpha).$$

(iv)   Take $n = 2$,   and plot graphs of $F(t)$ and $f(t)$.

(v)   Extend this problem to infinite n.

42.   Let X be a life time with mean μ,   variance σ^2 and
a d.f. F.   Let X_1 and X_2 be two independent life
times with the same distribution as X.   Consider

$$Z \;=\; \min(X_1, X_2).$$

Show that

$$\int_0^\infty F(x) F^c(x) \; dx \;=\; \mu - m \;>\; 0$$

where $m = E(Z)$.

43.   Consider the extreme life times M_+ and M_- defined in
Section 6 for i.i.d. life times, with the common life
time X.   Show analytically that always:

$$E(M_-) \;\leq\; E(X) \;\leq\; E(M_+).$$

44. Show that moments of the life time X with d.f. F
     are given by:

$$E(X^n) \;=\; n \int_0^\infty x^{n-1} F^c(x) \, dx, \qquad n = 1,2,\ldots \, .$$

45. Let f_1 and f_2 be two densities of (positive) life times,
     and let g be their convolution (as in Section 5).

   (i) Show that

$$\int_0^z x f_1(x) f_2(z - x) \, dx + \int_0^z y f_2(y) f_1(z - y) \, dy$$

$$= zg(z), \qquad z \geq 0.$$

   (ii) Show that

$$\int_0^\infty x f_1(x) \, dx + \int_0^\infty y f_2(y) \, dy = \int_0^\infty z g(z) \, dz.$$

46. Let X be an exponential life time with mean $1/\lambda$.

   (i) Show that for any two instants s and t such that
        $0 \leq s < t$ one always has

$$F(t + h) - F(t) < F(s + h) - F(s).$$

        Illustrate this inequality on the graph of den-
        sity f.

   (ii) Show that for a fixed $t \geq 0$, the conditional
         probability that $t < X \leq x$, given that $X > t$,
         has the form

$$\mathrm{pr}(t < X \leq x | X > t) = 1 - e^{-\lambda(x-t)}, \qquad x \geq t.$$

(iii)   Find points a < b such that the following relations hold:

$$pr(X \le a) + pr(X > b) = pr(a < X \le b),$$

$$pr(X > a) = pr(X \le b),$$

and verify that $pr(X \le a) = pr(X > b) = 1/4$, and
$\lambda(b - a) = \log 3$.

(iv)   Assuming the cost function of the form

$$\phi(x) = \begin{cases} 0 & \text{for } a < x < b \\ c > 0 & \text{otherwise} \end{cases}$$

(where c is a constant), show that the average
cost is c/2.

47.   Suppose that in some problem the hazard rate r is zero
as long as life stays below a threshold value b. At b,
the rate jumps to a fixed value λ and then stays constant. Thus, the hazard rate has the form

$$r(x) = \begin{cases} 0 & \text{for } 0 \le x \le b. \\ \lambda & \text{for } b < x < \infty. \end{cases}$$

(i)   Compute the hazard function R for all $x \ge 0$.
(ii)   Find the life time distribution function F for
all $x \ge 0$.
(iii)   Show that the mean life time is $b + 1/\lambda$.
(iv)   If the cost function is given by

$$\phi(x) = \begin{cases} 1/x & \text{for } 0 < x < b \\ c(x - b) & \text{for } b \le x < \infty, \ c > 0 \end{cases}$$

find the average cost.

(v)   Determine the conditional d.f. $K_t(h)$ of prolongation, assuming $t > b$.

48.   Suppose that in some industrial problem the hazard rate
r has the form

$$r(t) = \lambda/\sqrt{t} \qquad \text{for } t > 0, \ \lambda > 0.$$

(i)   Compute the hazard function R, and verify that
$R(x) \to \infty$ as $x \to \infty$.

(ii) Find the life time distribution function F, determined by R, for the life time X.

(iii) Find the mean life time $\mathbb{E}X$.

(iv) Show that in the present case $\mathbb{E}X = \mathbb{E}\sqrt{X}/\lambda$.

(v)   Sketch graphs of functions r, R, and F.

49.   Suppose that in some problem the hazard rate r has the
form

$$r(x) = \begin{cases} k/(a - x) & \text{for } 0 < x < a, \\ 0 & \text{otherwise,} \end{cases}$$

where $a > 0$ and a positive integer k are constant.

(i)   Compute the hazard function R for $x < a$, and
verify that $R(x) \to \infty$ as $x \to a$.

(ii) Find the life time distribution function F for
all $x \geq 0$.

(iii) Find the mean life time.

(iv) If the cost function is given by

$$\phi(x) = cx \qquad \text{for } 0 < x < a$$

(where $c > 0$), find the average cost.

50. It has been observed that the duration of a production process follows the uniform distribution for small values of time, say till a fixed duration L, and the exponential distribution for durations larger than L. Thus, the density of X has the form

$$f(x) = \begin{cases} (1 - c)/L & \text{for } 0 < x < L, \\ c\lambda e^{-\lambda(x-L)} & \text{for } L < x < \infty, \end{cases}$$

where c is a scaling factor such that $0 < c < 1$, and L and λ are positive.

   (i)   Find the distribution function F.

   (ii)   Plot graphs of f and of F.

   (iii) Show that the average duration is $\mathbb{E}X = (1 + c)L/2 + c/\lambda$.

   (iv)   What is the value of the hazard rate r(x) for $x > L$?

51. Let X be a life time with the uniform distribution over (0,L), and let Y be a life time with the exponential distribution with a parameter λ.

   Consider two cost functions ϕ and ψ defined respectively by, for X,

$$\phi(x) = \begin{cases} ce^{-\lambda x}, & 0 \le x \le L \\ 0, & x > L \end{cases}$$

   for Y:

$$\psi(y) = \begin{cases} c, & 0 \le y \le L \\ 0, & y > L \end{cases}$$

where $c > 0$.

Show that the average costs are equal:

$$\mathbb{E}\phi(X) = \mathbb{E}\psi(Y),$$

whenever $2\mathbb{E}X = \mathbb{E}Y$.

52. Let X be a life time with distribution given in Problem 5, but with $a = 1$ and $b > 0$.

   (i) Show that $\mathbb{E}X = \infty$.

   (ii) Consider the cost function defined by

$$\phi(x) = \begin{cases} x & \text{for } 0 \le x < k, \\ 0 & \text{for } x \ge k, \end{cases}$$

   where $k > 0$ is a constant. Show that the average cost satisfies

$$\mathbb{E}\phi(X) = b \log (1 + k/b) - k/(b + k) \to \infty$$
$$\text{as } k \to \infty.$$

53. Consider equipment whose life time X has the exponential distribution. Suppose that the cost of operation increases proportionally with the age of the equipment as long as X increases from 0 to a critical value a. At instant a, the cost drops to zero and again increases linearly with life time. Thus, the cost function has the form

$$\phi(x) = \begin{cases} cx & \text{for } 0 \le x < a \\ c(x - a) & \text{for } a \le x < \infty \end{cases}$$

for fixed $c > 0$. Find the average cost $\mathbb{E}\phi(X)$, and show that as the function of a, it has minimum when a equals the average life time.

54. Consider equipment whose life time X has uniform distribution over the interval $(0,L)$. Suppose that the

cost of operation is zero when the life X is larger
than some value a, but for X smaller than a it is
given by the quadratic function stated below.

Thus, the cost function has the form

$$\phi(x) = \begin{cases} c(L - x)x & \text{for } 0 \leq x \leq a, \\ 0 & \text{for } a < x \leq L, \end{cases}$$

where $0 \leq a \leq L$, $c > 0$.

Find the average cost $\mathbb{E}\phi(X)$, and sketch its graph as
a function of a. Verify that this cost increases with
a and that the graph has an inflection point at a = $\mathbb{E}X$.

55. Consider equipment containing two components whose
life times X and Y are i.i.d. with the common density
f. Find the average cost of operation in the two
separate cases described below.

  (i)   Suppose that the cost of operation of the
         system is constant, say a, when X > Y, and is
         b when X < Y (where a ≠ b). Then the average
         cost is (a + b)/2.

  (ii)   Suppose that the cost of operation of the
         system is a whenever X + Y < z and is b when-
         ever X + Y > z, where a ≠ b and z is a fixed
         positive threshold value. Then the average
         cost is an increasing function of z when a > b.

56. Let X and Y be two independent life times, each uniformly distributed over different intervals (a,a + L)
and (0,b), respectively, but with the same mean values
$\mathbb{E}X = \mathbb{E}Y$. (Here a, b, and L are positive constants.)

(i)   Show that

$$a + L \leq b$$

(ii)   Show that

$$pr(X \geq Y) = 1/2$$

57. Suppose that time X needed to write a letter has uni-
     form distribution over an interval (0,L) but the
     "preparation time" Y (needed to find a writing pad,
     envelope, address, and stamp) has exponential distribu-
     tion with a parameter λ.

     Assuming that X and Y are independent, find the
     probability that more time will be spent on preparation
     than on the actual writing. Verify that if both times
     have the same mean value, then the required probability
     is

$$(1 - e^{-2})/2$$

58. Suppose that time X needed to eat a pie and time Y
     needed to drink coffee are dependent in such a way
     that their joint density f(x,y) is constant, say k,
     over a triangle bounded by lines x = 0, y = 0, and
     x + y = a > 0, and is zero outside it.

     Show that the probability of ratio Y/X being larger
     than a fixed number z is

$$pr(Y/X > z) = 1/(1 + z),$$

     independently of a. Verify that this probability is
     actually x_0/a where (x_0, y_0) is the solution of the
     system of equations x + y = a and y = zx.

59. A designer can select either device A or device B.
     The life time of device A has exponential distribution
     and operates with cost function of the form cx^n (for
     $0 \leq x < \infty$), whereas the life time of device B has
     uniform distribution over the interval $(0,L)$ and
     operates with cost function of the form $c(L - x)^n$
     (for $0 \leq x \leq L$). In both cases n is a positive integer
     and c is a positive constant.

     Assuming that both life times have the same mean,
     show that device B has smaller average cost than
     device A, whenever $n \neq 1$.

60. Show that the standard normal density ϕ integrates to
     1 by evaluating the integral

     $$\int_0^\infty \phi(z) \ dz = 1/2$$

     using substitution $z^2 = 2t$ and the gamma function
     $\Gamma(1/2) = \sqrt{\pi}$ (see Appendix A). Compare with the method
     used in Section 9.

61. The gamma density of a life time is defined by

     $$f(x) = \frac{(\lambda x)^{\gamma-1}}{\Gamma(\gamma)} \ e^{-\lambda x} \ \lambda, \qquad x > 0,$$

     where $\lambda > 0$ and $\gamma > 0$ are constants and $\Gamma(\gamma)$ is the
     gamma function (see Appendix A).

     Calculate:

     (i) Laplace transform:

     $$f^*(\alpha) = [\lambda/(\lambda + \alpha)]^\gamma$$

(ii) moments of order k:

$$\mu_k = \lambda^{-k} \, \Gamma(\gamma + k)/\Gamma(\gamma)$$

(iii) mean and variance:

$$\mu = \gamma/\lambda, \qquad \sigma^2 = \gamma/\lambda^2.$$

Note: Gamma life times are of importance in many applications, especially when γ is a positive integer so $\Gamma(n) = (n - 1)!$. See, for example, Section 16. Note that the special case $\gamma = 1$ yields to exponential density. Problem 39 treats case $\gamma = 1/2$.

62. The beta density of a life time is defined by

$$f(x) = \frac{1}{B(p,q)} \, x^{p-1} \, (1 - x)^{q-1}, \qquad 0 \le x \le 1,$$

where $p > 0$ and $q > 0$ are constants and $B(p,q)$ is the beta function (defined in Appendix A).

Show that moments of order k are given by

$$\mu_k = \frac{B(k + p,q)}{B(p,q)}$$

and in particular the mean and variance are

$$\mu = p(p + q)^{-1},$$

$$\sigma^2 = pq \, (1 + p + q)^{-1}(p + q)^{-2}.$$

Note: Beta distribution generalizes the uniform distribution (corresponding to $p = q = 1$). In most applications p and q are integers (for example, Problem 3— with proper scaling—corresponds to $p = q = 2$).

63.   Let X and Y be two independent life times having the
      gamma density with the same parameter λ but with
      different parameter γ, say integers n and m, respec-
      tively.

      Show that the life times

$$W = X/(X + Y) \qquad \text{and} \qquad V = Y/(X + Y)$$

      have each beta distribution with parameters p = n, q =
      m and p = m, q = n, respectively. Hence,

$$\mathbb{E}W = n/(n + m), \qquad \mathbb{E}V = m/(n + m).$$

      <u>Note</u>: This is the extension of Example 4 in Section
      8.

64.   Let X and Y be two independent identically distributed
      life times with common density f. Show that propor-
      tions

$$W = X/(X + Y) \qquad \text{and} \qquad V = Y/(X + Y)$$

      have common distribution with density

$$g(t) = \int_0^\infty f(tz)f(z - tz)z \, dz, \qquad 0 < t < 1.$$

      (For expoenential f, see Example 4 in Section 8.)

65.   Let X be a life time with density f such that f(0) =
      0. Let Y by an exponential life time with mean μ.

      Suppose that there exists a decomposition

$$X = Y + Z,$$

      where Y and Z are independent. Let g be the density
      of Z. Show that g must have the form

$$g(x) = f(x) + \mu \, df(x)/dx, \qquad x \geq 0.$$

Is such a decomposition always possible? Find g when f has the form

$$f(x) = xe^{-x^2/2}, \qquad x \geq 0.$$

(<u>Note</u>: Determination of required g is known as de-convolution.)

2
Be Discreet with Discrete

We shall now discuss life times which assume values in
a countable set only (that is finite or denumerably infinite).
We shall call them "discrete life times." This may be for
example the waiting time measured in fixed units, like 1,
2, 3 and so on, hours, days, etc. We can also make observations at specific instants of time, say at noon every day.

In this category we may incorporate other variables not
usually designated as life times, like, for example, a number of objects in some collection. The number of people
waiting in a line, the number of patients in a hospital, the
number of engaged machines in a shop, and many other examples
of this kind, are typical illustrations.

The main difference from the previously considered situation is that we shall not use integration (or differentiation), but rather summation (finite or infinite). In this
respect our discussion will be simpler, although some of our
sums may be rather strange.

This Chapter deals with three main discrete distributions, namely binomial, geometric and Poisson -- they corre-

spond to three well known series from calculus. On the application side, we are in the heart of Probability Theory.
Bernoulli trials (Sections 11-13) go back to James Bernoulli,
a 17th century Swiss mathematician; and the Poisson distribution originated with Simeon D. Poisson, a French mathematician working in the first half of the 19th century. It
is amazing how many modern applications fit well into these
venerable formulae.

We shall see here many applications to diverse fields,
ranging from simple and amusing to more complicated and serious examples. We shall also hint at some modern develop-
ments which lie behind surface (see Section 15, and some
Problems).

Section 11: Bernoulli Trials

11.1.

We shall begin with a rather important case of Bernoulli
trials which find an extremely large number of applications.
Suppose that we perform an "experiment" which may result
only in two outcomes. This of course need not be a physical
experiment, but any observation, situation, or action in
which only two aspects are of interest. For example, taking
an exam one may pass or fail, a signal may be on or off, in
a game one can win or lose, and so on; we need not mention
the well known coin tossing! For convenience, any such
attempt or experiment will be referred to as a <u>trial</u>. Its
outcomes are conveniently called a success (S) and a failure

(F); what is S and what is F is immaterial (one man's success is another man's failure).

We shall consider an event S (that success occured) and an event F (the failure took place), and in agreement with our procedures (Section 1) associate probabilities with these events:

$$Pr(S) = p, \qquad pr(F) = q,$$

where $p + q = 1,$ $0 \le p \le 1,$ $0 \le q \le 1.$

The condition $p + q = 1$ is in agreement with the fact that S and F are two complementary events.

Now, we consider an experiment which consists of repeating such trials n times. This may be, for example, n exams to be taken, n machines in a shop, n sources sending signals, etc. Each of these n trials may result either in a success or in a failure. The big problem is to find the probability that n trials will result in exactly j successes, where j ranges from 0 to n. If n is a number of chairs in the room, and success means that a chair is occupied (i.e., there is a customer in the room), then our problem is to have the expression for the probability that there will be exactly j chairs occupied; if j = 0, all chairs are free, if j = n, all chairs are taken.

To be more specific, we must impose further assumptions which will specify <u>Bernoulli trials</u>:

i)   all trials have the same pr. p of resulting in
     success S,

ii)   all trials are independent of each other.

   The assumption of independence is essential, and it
simply means that, say, the pr. of two successes SS is
simply p^2, pr. of FF is q^2, pr. of SF is pq, and
so on (this is analogous to notion of independence discussed
in Section 5).

   Let X be a "life time" (i.e., a random variable)
representing a number of successes in n trials. We shall
write for its distribution, that is the probability:

$$pr(X = j) = P(j), \qquad j = 0,1,\ldots,n.$$

Since the events of getting 0 successes, 1 success, ...,
j successes are necessarilly exclusive the pr.'s add to
unity:

$$\sum_{j=0}^{n} P(j) \;=\; P(0) + P(1) + \ldots + P(n) \;=\; 1.$$

It is our task to find the formula for P(j). For this purpose we shall examine events (X = j) successively for each
j, as in the following table. Remember, there are n
trials and we look for all possible arrangements of successes
S.

X = j	sequence of length n	pr. of each sequence	$P(j) = pr(X=j)$
$j = 0$	FFF........F	q^n	q^n
$j = 1$	SF.........F F.....S....F F.........FS	pq^{n-1} each	npq^{n-1}
j	F..S..FS..F..S j S's $n-j$ F's	$p^j q^{n-j}$ each	$\binom{n}{j} p^j q^{n-j}$
$j = n$	SSS........S	p^n	p^n

Indeed, if in a string of n trials there is exactly j S's, then there must be exactly $n-j$ F's; the pr. of getting j S's is p^j, by independence; the pr. of getting $n-j$ F's is q^{n-j}; hence, the pr. of a sequence with j successes is (again by independence) $p^j q^{n-j}$.

However, j successes are distributed among n places -- the number of selections of j places out of n, that is a number of sequences of length n carrying j S's and $n-j$ F's, is given by the binomial coefficient

$$\binom{n}{j} = \frac{n!}{(n-j)!j!} ,$$

where $!$ denotes factorial:

$$k! = 1 \cdot 2 \cdots (k-1)k = k \cdot (k-1)!$$

with $0! = 1$.

Since each of such sequences carrying j S's is different (sequences being exclusive), the pr.'s add, and consequently the pr. of the composite event $(X = j)$ is

$$P(j) \;=\; \binom{n}{j} \, p^j (1-p)^{n-j}, \qquad\qquad j = 0,1,\ldots,n.$$

This expression is known as the <u>binomial</u> <u>distribution</u>, with parameters n,p. Observe that it agrees with values of j in the table.

11.2.

We shall look closely at the above expression for $P(j)$. First of all recall a few properties of the binomials coefficients:

$$\binom{n}{j} = \binom{n}{n-j}, \qquad \binom{n}{j} = \frac{n}{j}\binom{n-1}{j-1}, \qquad \binom{n}{j} + \binom{n}{j-1} = \binom{n+1}{j}.$$

For small values of n, the binomial coefficients are easily obtained from the so called Pascal triangle, as it is well known.

The name "binomial" comes from the binomial expansion:

$$(a+b)^n \;=\; \sum_{j=0}^{n} \binom{n}{j} \, a^j b^{n-j}$$

for any real a and b.

Taking $a = p$ and $b = q$, it follows immediately that $\sum_{j=0}^{n} P(j) = (p+q)^n = 1$.

Letting $a = b = 1$, one finds that the sum of all binomial coefficients is: $\sum_{j=0}^{n} \binom{n}{j} = 2^n$. Letting $b = 1$ and $a = -1$, one has:

$$\sum_{j=0}^{n} \binom{n}{j}(-1)^j = 0.$$

Consider now the ratio:

$$\frac{P(j)}{P(j+1)} = \frac{j+1}{n-j}\frac{q}{p}.$$

It follows that according to values n and p, the following possibilities may occur: $P(j)$ as a function of j may increase, may decrease, or may increase first and then decrease.

The expected number of successes $E(X)$ is defined by

$$E(X) = \sum_{j=0}^{n} j \cdot P(j).$$

Similarly, the expected cost (cf. Section 7) is

$$E\varphi(X) = \sum_{j=0}^{n} \varphi(j)\ P(j).$$

It can be shown that

$$\boxed{E(x) = np \quad \text{and} \quad \text{var}(X) = npq.}$$

Indeed:

$$E(X) = \sum_{j=0}^{n} j\binom{n}{j}p^j q^{n-j} = \sum_{j=0}^{n} j\,\frac{n!}{j!(n-j)!}\,p^j q^{n-j}$$

$$= n\sum_{j=1}^{n}\frac{(n-1)!}{(j-1)!(n-j)!}\,p^j q^{n-j} = np\sum_{j=1}^{n}\binom{n-1}{j-1}p^{j-1}q^{n-j}$$

$$= np\sum_{k=0}^{n-1}\binom{n-1}{k}p^k q^{n-1-k} = np \qquad (k = j-1).$$

In order to find var X, it is more convenient to compute EX(X-1) first:

$$EX(X-1) = \sum_{j=0}^{n} j(j-1) \frac{n!}{j!(n-j)!} p^j q^{n-j}$$

$$= n(n-1) \sum_{j=2}^{n} \frac{(n-2)!}{(j-2)!(n-j)!} p^j q^{n-j} = n(n-1)p^2$$

so $EX^2 = EX(X-1) + EX = n(n-1)p^2 + np = n^2 p^2 + npq.$
Hence var X $= EX^2 - (EX)^2 = n^2 p^2 + npq - (np)^2 = npq.$

****

Recall that $P(j)$ is the pr. of exactly j successes. In order to find the pr. of j or less successes, one has

$$pr(X \le j) = \sum_{i=0}^{j} P(i) = P(0) + P(1) + \dots + P(j).$$

This is clearly an increasing function of j. Obviously, $pr(X \le n) = 1.$ Similarly, the complementary pr. of more than j successes is

$$pr(X > j) = \sum_{i=j+1}^{n} P(i) = P(j+1) + \dots + P(n).$$

This is analogous to the d.f. F in the continuous case, except that now the argument is discrete:

$$\boxed{F(j) = \sum_{i=0}^{j} \binom{n}{i} p^i q^{n-i}, \quad 0 \le j \le n}$$

Note: Special cases of interest:

(i) $p = q = 1/2;$ $P(j) = \binom{n}{j} \dfrac{1}{2^n}$.

(ii) $p = 1;$ $P(n) = 1,$ $P(j) = 0,$ $j \neq n.$

(iii) $p = 0;$ $P(0) = 1,$ $P(j) = 0,$ $j \neq 0.$

11.3.

There is another approach to Bernoulli trials which is
very instructive. Let Z be a r.v. which assumes values 0
and 1, only, with probabilities:

$$pr(Z = 0) = q, \qquad pr(Z = 1) = p.$$

Hence, $E(Z) = p$ and $var(Z) = pq.$ Clearly, Z represents
the result of a trial (1 stands for success S and 0 for
failure F). Hence, r.v.'s $Z_1, Z_2, \ldots, Z_n$ which are i.i.d.
(independent identically distributed) with common distribution being that of Z, represent results of consecutive
n trials. The number of successes in n trials which we
now denote by X_n (to stress dependence on n) is therefore:

$$X_n = Z_1 + \ldots + Z_n.$$

As noted in Section 5 and also in Section 10.2, the mean and
variance of X_n are simply:

$$E(X_n) = np, \qquad var(X_n) = npq,$$

in agreement with our earlier calculations. Indeed, our
evaluation of P(j) is just the determination of distribution of X_n, in the manner of combination of life times as
discussed in Section 5.

Let us now consider the sample mean of successes in n
trials:

$$X_n/n \;\; = \;\; (Z_1 + \ldots + Z_n)/n.$$

Clearly, X_n/n has the same mean as Z (although
these r.v.'s are different), but its variance is $\mathrm{var}(X_n)/n^2$
and will decrease with n:

$$E(X_n/n) = p, \qquad \mathrm{var}(X_n/n) = pq/n.$$

By the Chebyshev inequality from Section 10.2 we have:

$$\mathrm{pr}(\,|X_n/n - p| > \varepsilon) \;\; \le \;\; pq/(n\varepsilon^2), \qquad \varepsilon > 0.$$

As $n \to \infty$, the bound goes to 0, so the indicated pr. tends
to zero. We thus obtained the Bernoulli Law of Large Numbers
which asserts that (with limit in probability, as in Section
10):

$$\lim_{n\to\infty} \frac{Z_1 + \ldots + Z_n}{n} \;\; = \;\; p.$$

Thus, as n becomes large the sample mean X_n/n for
Bernoulli trials converges (in probability) to the "popula-

tion mean" p (which coincides here with the pr. of a success).

This result contains a mathematical counterpart of the
intuitive notion of probability as frequency. The sample
mean X_n/n --actually its observable values -- yield the
frequency of occurrence of a success, and it tends to p
when the number of observations n becomes large. Compare
this statement with our earlier remarks on the frequency
approach in Section 1.

Section 12: Applications of Binomial

The binomial distribution discussed in the previous
section occurs frequently in applications, and this section
lists some illustrative examples.

12.1. Sampling with replacement

Suppose that there are M objects (coins, balls,
machines, cars) of which m are of a special kind in which
we are interested; label them as "good" objects.

An object chosen "at random," inspected and its kind
noted, and then it is returned. The procedure is repeated
n times (where, of course n may be larger than M); in
other words, a sample of size n is drawn with replacements
from a population of size M. What is the pr. that this
sampling will produce exactly j "good" objects?

Clearly, we have here the situation of n Bernoulli
trials, "at random" meaning that each object has equal chance

to be chosen, so p = m/M. Hence, the required pr. is:

$$P(j) = \binom{n}{j} (m/M)^j (1 - m/M)^{n-j}, \quad j = 0,1,\ldots,n$$

and the average number of good choices is nm/M.

As an example, what is the pr. that in a sequence of 7 digits (as in a telephone number) there will be at least two digits 5 ? Since there are 10 digits, M = 10, and 5 being one of them m = 1, so p = 1/10 obviously. Since n = 7, the required pr. is:

$$\sum_{j=2}^{7} P(j) = 1 - P(0) - P(1)$$

$$= 1 - (9/10)^7 - 7(9/10)^6 (1/10) = 0.17.$$

In this example the mean is np = 7/10 and variance is npq = 63/100. To compute the above expression we may use Gaussian approximation as in Section 9.

12.2. Overselling of tickets

An airline knows that on the average r of the people making reservations on a certain flight will not show up. Consequently, the policy is to sell t tickets for a flight that can only carry s passengers (t > s).

What is the pr. that there will be a seat available for every passenger that shows up? Let X be the number of passengers showing up; clearly, X assumes values from 0 to t. Hence, EX = t - r, and therefore p = 1 - r/t.

The pr. of exactly j passengers showing up is the binomial

$$P(j) = \binom{t}{j} p^j (1-p)^{t-j}.$$

Consequently, the required pr. is:

$$pr(X \leq s) = \sum_{j=0}^{s} P(j).$$

In the special case when only one seat is oversold, then this pr. is $1 - (1 - r/t)^t \approx 1 - e^{-r}$, as can be seen by putting $s = t - 1$.

As an example, take $r = 10$, $t = 100$, and $s = 95$. Then $p = 0.9$. The mean is $tp = 90$ and variance $tp(1-p) = 9$. Hence, the Gaussian approximation yields:

$$pr(0 \leq X \leq 95) = \Phi(\frac{95-90}{3}) - \Phi(\frac{0-90}{3}) = \Phi(\frac{5}{3}) - \Phi(-30)$$

$$\approx \Phi(\frac{5}{3}) = 0.952.$$

In another special case suppose that the average number of passengers showing up equals the number of seats; that is $t - r = s$. Then, $p = s/t$ and variance is $s(1 - s/t)$. Hence the Gaussian approximation yields:

$$pr(0 \leq X \leq s) = \Phi(\frac{s-s}{\sigma}) - \Phi(\frac{-s}{\sigma}) \approx \Phi(0) = \frac{1}{2}$$

which looks rather bad for a passenger.

12.3. <u>ESP</u>

An individual claims to have extrasensory perception (ESP). As a test, a fair coin is tossed n times and he is asked to predict in advance the outcome.

Our individual gets k out of n correct. What is the pr. that he would have done this well if he had no ESP?

Here again we have Bernoulli trials with $p = \frac{1}{2}$. Let X be a number of correct answers (i.e., proper matches). Then the required pr. is

$$Pr(X \geq k) \quad = \quad \sum_{j=k}^{n} \binom{n}{j}(1/2)^n.$$

Thus, if getting k or more out of n correctly is an event of small pr., then this would indicate that the individual has ESP.

As an example, suppose that n = 100, so the mean and variance are respectively np = 50, npq = 25, and the Gaussian approximation is:

$$Pr(X \geq k) \quad = \quad 1 - \Phi((k-50)/5)$$

and the numerical results are: for

$$k = 50 \quad \text{the pr. is} \quad 0.500$$
$$k = 55 \quad \text{the pr. is} \quad 0.136$$
$$k = 60 \quad \text{the pr. is} \quad 0.023.$$

12.4. <u>Telephone calls</u>

Consider a group of n devices which originate calls (telephone subscribers, lines, etc). The pr. that a device

is busy (i.e., carries a call) is   p.   In Telephony $p = cT$
where   c   is the calling rate per unit of time (i.e. the
inverse of the mean time between two consecutive calls
originated by a device) and   T   is the average duration of
a call; clearly $p < 1$.   The pr. that exactly   j   devices
out of   n   carry calls is given by binomial   P(j)   with
parameters   n, p.   The average number of busy devices   np
is called traffic.   The pr. that all devices are engaged
is clearly p^n.

   Suppose now that these   n   devices have access to   s
lines in some other group, where $s < n$.   Thus only   s
originated calls can go through, the other being blocked.
We would like to have the pr. of   j   calls going through,
i.e., not being blocked.   This is expressed by the conditional pr.

$$Q(j) \;=\; pr(X = j \mid X \leq s), \quad j = 0,1,\ldots,s.$$

This can be evaluated (as in Section 3) to give

$$Pr(X = j \mid X \leq s) \;=\; \frac{pr[(X = j) \cap (X \leq s)]}{pr(X \leq s)} \;=\; \frac{pr(X = j)}{pr(X \leq s)}$$

$$=\; \frac{\binom{n}{j} p^j q^{n-j}}{\sum\limits_{i=0}^{s} \binom{n}{i} p^i q^{n-i}} \;=\; \frac{\binom{n}{j} \left(\frac{p}{q}\right)^j}{\sum\limits_{i=0}^{s} \binom{n}{i} \left(\frac{p}{q}\right)^i} \;.$$

   The above expression   Q   is known as the truncated
binomial distribution.   Of special interest is that the pr.

of blocking ,defined as the pr. that all s   lines are occupied:

$$Q(s) \;=\; \frac{\binom{n}{s}(\frac{p}{q})^{s}}{\sum\limits_{i=0}^{s} \binom{n}{i}(\frac{p}{q})^{i}} \;.$$

12.5. Cost

Suppose that the cost function φ has the form:

$$\varphi(j) = C > 0, \qquad \text{for} \quad j = 1,\ldots,n-1$$
$$= 0 \qquad\qquad \text{for} \quad j = 0,\, n.$$

Then the average cost is:

$$E\varphi(X) \;=\; C \sum\limits_{j=1}^{n-1} P(j) \;=\; C(1 - P(0) - P(n)) \;=\; C(1 - p^{n} - q^{n}).$$

****

Section 13:   Geometric Waiting Time

Consider again Bernoulli trials, but instead of fixing
the number of trials, suppose that we continue trials till
the first success appears. This may happen of course at
the first attempt, or we may perform a large number of trials
-- that is we obtain a succession of failures -- and then
the first success appears. In contrast with the previous
section, the number of trials is now a random variable and
actually the discrete life time. Typical examples are: the

length of a "lucky" string of wins (in a game, in passing examinations), collecting of coupons, time needed to the first occurrence of some event, waiting time, etc.

We consider now the Bernoulli trials with $p = pr(S)$ and $q = pr(F)$, with $p + q = 1$. Let N be a life time of period of F's until the first S, that is the waiting time until the first S. Clearly, N assumes as its value all positive integers: $1,2,3,\ldots$. Denote the distribution of N by:

$$P(n) = pr(N = n), \quad n = 1,2,\ldots .$$

This is the pr. that the first success occurs exactly at the nth trial. To find $P(n)$ consider events $(N = n)$ for each n, as in the following table:

n	sequence	P(n)	(by independence)
1	S	p	
2	FS	pq	
3	FFS	pq^2	
...	...	...	
n	F...FS	pq^{n-1}	(here n-1 F's followed by one S)

Consequently the required distribution is given by:

$$\boxed{P(n) = pq^{n-1}, \quad n = 1,2,\ldots} .$$

Observe that $P(n+1)/P(n) = q$, so P decreases with increasing n. The distribution P is called the geometric distri-

bution, because of its relation to the geometric series:

$$\text{for } |\alpha| < 1: \quad \sum_{i=0}^{\infty} \alpha^i = \frac{1}{1-\alpha} \; .$$

Writing $q = 1 - \alpha$, it follows that:

$$\sum_{n=1}^{\infty} P(n) = 1.$$

The life time N assumes infinitely many values, but the pr. of waiting infinitely long is zero, because $P(n) \to 0$, as $n \to \infty$.

The pr. of waiting more than n is clearly:

$$pr(N > n) = \sum_{k=n+1}^{\infty} pq^{k-1} = pq^n \sum_{k=n+1}^{\infty} q^{k-n-1} = pq^n/p = q^n.$$

Consequently, the pr. of prolongation of the life time by additional h units is (as in Section 3):

$$pr(N > n+h \mid N > n) = pr(N > n+h)/pr(N > n) = q^{n+h}/q^n = q^h.$$

This pr. does not depend on the duration of life, but only on the amount h of prolongation. Thus, the geometric distribution is memoryless; it is the analogue of the n.e.d. in the continuous case.

The average waiting time $E(N)$ is defined as always by $E(N) = \sum_{n=1}^{\infty} nP(n)$. Direct computation is rather cumbersome: first display the above sum in the following way

$$\sum_{n=1}^{\infty} nP(n) \;=\; \begin{cases} p \\ pq + pq \\ pq^2 + pq^2 + pq^2 \\ \qquad \cdots \\ pq^n + \cdots + pq^n \qquad (n{+}1 \text{ terms}) \\ \qquad \cdots \end{cases}$$

and then sum vertically to get

$$\sum_{k=0}^{\infty} p\left(\sum_{i=k}^{\infty} q^i\right) \;=\; p\sum_{k=0}^{\infty} q^k/p \;=\; \sum_{k=0}^{\infty} q^k \;=\; 1/p.$$

Thus

$$\boxed{E(N) \;=\; 1/p} \;;$$

it can be shown that

$$\boxed{\text{var}(N) \;=\; qp^{-2}} \;.$$

Note that strictly speaking one should have $0 \le q < 1$. Indeed, for $q = 0$, obviously

$$P(1) = 1, \quad P(n) = 0 \quad \text{for} \quad n \ne 1, \quad E(N) = 1$$

because then necessarily the first trial results in success. On the other hand, for $q = 1$,

$$P(n) = 0 \quad \text{for all} \quad n, \quad \text{and} \quad \text{pr}(N > n) = 1 \quad \text{for all} \quad n,$$

$$E(N) = \infty.$$

One can say that the first success occurs at infinity, that it is never in finite time.

In the intermediary case when $p = q = 1/2$, obviously

$$P(n) = (1/2)^n \quad \text{and} \quad E(N) = 2.$$

<u>Example 1</u>: Suppose that an applicant for a driver's license has 60% chance to pass the test on any given try. Then the pr. that the applicant will eventually get the license on the third try is

$$P(3) = 0.6 \times 0.4^2 = 0.096.$$

<u>Example 2</u>: What is the pr. that in tossing a coin, the first head will appear at the 10th toss? Here $p = 1/2$, so

$$P(10) = (1/2)^{10}$$

<u>Example 3</u>: Let E be the set of all (positive) even integers, and let 0 be the set of all (positive) odd integers. Suppose that positive integers are distributed according to the geometric distribution. Then

$$pr(0) = \sum_{k=0}^{\infty} pq^{2k+1-1} = p \sum_{k=0}^{\infty} q^{2k} = \frac{p}{1-q^2} = \frac{1}{1+q}$$

because for odd $n = 2k + 1$. Therefore,

$$Pr(E) = \frac{q}{1+q}.$$

Note that the sets of even and of odd integers have different pr.'s, despite the intuitive feelings that these sets are "similar." In particular, for $p = q = \frac{1}{2}$, one has

$$pr(O) = \frac{2}{3} \quad \text{and} \quad pr(E) = \frac{1}{3} .$$

Example 4: Assuming the geometric distribution on positive integers, find the pr. that an integer is even when it is divisible by 3. Clearly, this is the conditional pr.:

$$pr(\text{by } 2 \mid \text{by } 3) = \frac{pr(\text{by } 6)}{pr(\text{by } 3)} .$$

Hence, using the properties of the geometric series:

$$pr(\text{by } 6) = p \sum_{k=1}^{\infty} q^{6k-1} = \frac{p}{q} q^6 \sum_{k=1}^{\infty} (q^6)^{k-1} = \frac{pq^5}{1 - q^6}$$

$$pr(\text{by } 3) = p \sum_{k=1}^{\infty} q^{3k-1} = \frac{p}{q} q^3 \sum_{k=1}^{\infty} (q^3)^{k-1} = \frac{pq^2}{1 - q^3} .$$

Hence the required pr. is:

$$pr(\text{by } 2 \mid \text{by } 3) = q^3 \frac{1-q^3}{1-q^6} = \boxed{\frac{q^3}{1+q^3}} .$$

Note that this pr. is always smaller than the pr. that the integer is even (found in Example 3). Indeed:

$$\text{for } q > 0: \quad \frac{q^3}{1+q^3} < \frac{q}{1+q} \Rightarrow q^3 + q^4 < q + q^4 \Rightarrow q^2 < 1.$$

Note that for $q = \frac{1}{2}$, these pr.'s are respectively

$\frac{1}{9} < \frac{1}{3}$.

Section 14: Poisson Distribution

Let us return to Bernoulli trials and to the binomial distribution in Section 11. In many applications the number of trials n is large, and computation of P(j) may be cumbersome. It is therefore of interest to see if some approximation could be found (apart from Gaussian), when n becomes large. Letting n go to infinity would produce a meaningless result, so this limit must be taken with some restriction to obtain a reasonable answer. Indeed, in many cases one encounters the situation that when the number of trials n **becomes** very large, at the same time the pr. of success p becomes very small (the so called rare events), yet the average np remains practically the same. This suggests checking the following limit:

$\lim_{n \to \infty} P(j)$ **with p → 0 in such a way that**
$np = \mu$ **is constant.**

This turns out to be the right approach. To compute the limit the use must be made of the Stirling formula for approximation to the factorial:

$$n! \approx \sqrt{2\pi}\, n^n e^{-n}$$

and the following limit from calculus:

$$\lim_{n\to\infty}(1 - \frac{a}{n})^n = e^{-a}, \quad \text{for any } a.$$

Now we can write:

$$\binom{n}{j} p^j(1-p)^{n-j} = \frac{n!}{j!(n-j)!} p^j(1-p)^{n-j}$$

$$= \frac{n^n e^{-n}}{j!(n-j)^{n-j}e^{-n+j}} (\frac{\mu}{n})^j (1 - \frac{\mu}{n})^{n-j}$$

$$= \frac{\mu^j}{j!} (\frac{n}{n-j})^{n-j} (1 - \frac{\mu}{n})^{n-j} e^{-j} \to \frac{\mu^j}{j!} e^{-\mu}.$$

This yields the result:

$$P(j) = \frac{\mu^j}{j!} e^{-\mu}, \quad j = 0,1,\ldots$$

This is known as the <u>Poisson</u> <u>distribution</u>. Observe that j ranges from 0 to infinity (as n is now infinite). P(j) gives the pr. of exactly j successes in the infinite number of trials. Here the Poisson distribution has been obtained as the approximation to the binomial. We shall see later that it stands on its own merits; and can be derived independently of the binomial.

We must first verify that

$$\sum_{j=0}^{\infty} P(j) = 1.$$

This follows immediately from the series representation for e:

$$\sum_{j=0}^{\infty} a^j/j! \ = \ e^a.$$

From $P(j+1)/P(j) = \mu/(j+1)$ it is easy to see that $P(j)$ increases and then decreases with j, when $\mu > 1$, the maximum occurring for j being the integer between $\mu - 1$ and μ; for $\mu \leq 1$, $P(j)$ decreases with j.

It is very easy to find the mean of a random variable with the Poisson distribution:

$$E(X) \ = \ \sum_{j=0}^{\infty} jP(j) \ = \ \sum_{j=1}^{\infty} j \, \frac{\mu^j}{j!} \, e^{-\mu}$$

$$= \ \mu e^{-\mu} \sum_{j=1}^{\infty} \mu^{j-1}/(j-1)! \ = \ \mu e^{-\mu} e^{\mu} \ = \ \mu.$$

To get variance, consider first $EX(X-1)$; by the same argument

$$EX(X-1) \ = \ \sum_{j=2}^{\infty} j(j-1) \, \frac{\mu^j}{j!} \, e^{-\mu} \ = \ \mu^2 e^{-\mu} \sum_{j=2}^{\infty} \mu^{j-2}/(j-2)!$$

$$= \ \mu^2 e^{-\mu} e^{\mu} \ = \ \mu^2.$$

Hence $EX^2 = \mu^2 + \mu$, so $\text{var } X = EX^2 - (EX)^2 = \mu^2 + \mu - \mu^2 = \mu$. Thus, we have a very interesting property of the Poisson distribution:

$$\boxed{EX \ = \ \text{var } X \ = \ \mu} \ ,$$

where μ is the parameter occurring in the formula $P(j)$.

When it is required to find the distribution function
of X, denoted F(k) with k restricted to non-negative
integers, one has simply to add:

$$F(k) = pr(X \le k) = \sum_{j=0}^{k} P(j) = \sum_{j=0}^{k} \frac{\mu^j}{j!} e^{-\mu}.$$

The values of P(j) and F(k) are tabulated for given μ.

As an exercise in calculus, we can find an interesting
expression for F(k) by treating F as a function of μ
(for constant k). Differentiation with respect to μ yields

$$\frac{dF}{d\mu} = \sum_{j=1}^{k} \frac{\mu^{j-1}}{(j-1)!} e^{-\mu} - \sum_{j=0}^{k} \frac{\mu^j}{j!} e^{-\mu} = -\frac{\mu^k}{k!} e^{-\mu}.$$

Then, integration, with initial condition F = 1 for μ = 0,
yields

$$F(k) = 1 - \frac{1}{k!} \int_{0}^{\mu} t^k e^{-t}\, dt = \frac{1}{k!} \int_{\mu}^{\infty} t^k e^{-t}\, dt.$$

****

Example 1: A book of 200 pages contains 100 misprints.
Find the pr. that a given page contains at least 2 misprints. The average number of misprints per page is
μ = 100/200 = 0.5, so

$$pr(X \ge 2) = 1 - F(1) = 1 - 0.9098 = 0.0902.$$

Example 2: The average number of calls received by a switch
board during a fixed period is 5. What is the pr. that 6

or less calls will be received? Clearly F(6) = 0.7622

(for $\mu = 5$).

Example 3: Let E be the set of all (nonnegative) even
integers, and let 0 be the set of all (nonnegative) odd
integers. Suppose that nonnegative integers are distributed
according to the Poisson distribution. Then

$$pr(E) = \sum_{i=0}^{\infty} \frac{\mu^{2i}}{(2i)!} e^{-\mu} = (1 + \frac{\mu^2}{2!} + \frac{\mu^4}{4!} + \ldots)e^{-\mu}$$

$$= e^{-\mu} \cosh \mu,$$

using the series expansion for cosh, and the relation
$\cosh \mu + \sinh \mu = e^{\mu}$. Hence:

$$pr(0) = e^{-\mu} \sinh \mu.$$

Example 4: Colorblindness appears in 1% of people in a
certain population. How large must a random sample (with
replacement) be if the pr. of its containing a colorblind
person is to be 0.95 or more.

The pr. of at least one is $1 - P(0) = 1 - e^{-\mu} \geq 0.95$
but $\mu = np = n/1000$, so

$$e^{-\mu} \leq 0.05, \quad \text{or} \quad n \geq 300.$$

Example 5: What is the pr. that in a class of 110 students
exactly 2 will have birthdays today? These are Bernoulli

trials with $p = 1/365$ and $n = 110$, so $\mu \approx 0.3$, and $P(2) = 0.033$.

Example 6: Suppose that only s lines out of a (infinite) group can be connected to the second stage in a telephone system. Find the pr. that exactly j lines are occupied, given that this number does not exceed s.

   This is clearly

$$\text{pr}(X = j \mid X \leq s) = \frac{\text{pr}(X = j)}{\text{pr}(X \leq s)} = \frac{P(j)}{F(s)} = \frac{\frac{\mu^j}{j!}}{\sum_{i=0}^{s} \frac{\mu^i}{i!}}$$

$$\text{for } j = 0, 1, \ldots, s.$$

This is known as the Erlang distribution (or truncated Poisson): its value for $j = s$ is the famous Erlang formula for pr. of blocking.

Section 15: Accidents Just Happen

   In consideration of Bernoulli trials (Section 11) the number of trials was fixed, but the number of successes was a random variable. In the discussion of the waiting time (Section 13), the number of trials till the first success was the random variable. In many problems, however, both the number of trials and the number of successes are random variables. We shall now consider such a situation, and we will be interested in finding the distribution of successes.

Consider a fluctuating population of some objects of a
specified kind (say bacteria in a culture, accidents in a
city, cars in some region, houses in a community, eggs in a
shopping basket) and assume that the number of objects N
has Poisson distribution with mean λ. That is, the $\mathrm{pr}(N = n)$
$= Q(n)$ of having exactly n objects is given by

$$Q(n) \;=\; \frac{\lambda^n}{n!}\, e^{-\lambda}, \qquad n = 0,1,\dots\;.$$

Suppose that these objects may suffer some misfortune like
damage, loss etc., which for convenience we shall call an
<u>accident</u>. (For example, death of a bacteria, fatal accident
in a city, broken car, house fire, rotten egg). Assume that
accidents occur independently, each with pr. p. That is,
we have here Bernoulli trials, and the pr. of exactly j
accidents among n objects, denoted by $P(j|n)$, is now

$$P(j\,|n) \;=\; \binom{n}{j} p^j (1-p)^{n-j}, \qquad j = 0,1,\dots,n.$$

We will be interested in the pr. that there will be
exactly j accidents, irrespective of the number of objects
in the population. Denote by S the number of accidents;
we wish to find

$$\mathrm{pr}(S = j) = P(j), \qquad j = 0,1,\dots\;.$$

Clearly, we are looking for the distribution of dead bacteria,

fatal accidents, broken cars, fires, rotten eggs, as the case
may be.

From the above description it is clear that we have
"successes" (i.e., accidents) in   n   Bernoulli trials, so in
fact the conditional pr.:

$$pr(S = j \mid N = n) = P(j \mid n).$$

By properties of conditional pr. (see Section 3), the joint
pr. is

$$pr(S = j, N = n) = P(j \mid n)Q(n).$$

Taking summation over all   n,   we obtain the marginal distribution:

$$pr(S = j) = \sum_n P(j \mid n)Q(n) .$$

This is the required distribution $P(j)$.

It should be emphasized that here the binomial distri-
bution $P(j \mid n)$ and the Poisson distribution $Q(n)$ are quite
separate and not related to each other.   (That's why mean
has been denoted by λ in order to avoid confusion with
$\mu = np$ from Section 14).

The explicit calculation of $P(j)$ proceed easily as
follows: Observe first that summation in the above expression is actually from $n = j$,   becuase it is impossible to

have   n < j.   Note that this is automatically taken care of
by the fact that

$$\binom{n}{j} = 0 \quad \text{for} \quad n < j.$$

Hence

$$\boxed{P(j)} = \sum_{n=j}^{\infty} \binom{n}{j} p^j (1-p)^{n-j} \frac{\lambda^n}{n!} e^{-\lambda} = \frac{p^j \lambda^j}{j!} e^{-\lambda} \sum_{n=j}^{\infty} \frac{(1-p)^{n-j} \lambda^{n-j}}{(n-j)!}$$

$$= \frac{p^j \lambda^j}{j!} e^{-\lambda} e^{(1-p)\lambda} = \boxed{\frac{(\lambda p)^j}{j!} e^{-\lambda p}} \quad , \quad j = 0,1,\dots \; .$$

Thus we obtain the very interesting result that the
number of accidents is also Poisson distributed, but with
mean λp.   So

$$E(S) = \lambda p = \text{var } S.$$

The effect of individual accidents is reflected in appearance
of factor   p,   which reduces the mean λ to the new mean
λp.   Recall that if   Z   is a r.v. equal to one if a success
occurs, and to zero if not, then $EZ = p$.   Hence we can
write

$$ES = EN \cdot EZ$$

which has an intuitive meaning: the average number of accidents equals the average number of objects times the average
value of accident proneness.

****

As another illustration, suppose that the population distribution $Q(n)$ is geometric. As in the present situation n ranges from zero, the geometric distribution is taken of the form:

$$Q(n) = (1-a)a^n, \quad n = 0,1,\ldots, \quad 0 < a < 1.$$

Here $EN = a(1-a)^{-1}$.

Hence, as before:

$$P(j) = \sum_{n=j}^{\infty} \binom{n}{j} p^j (1-p)^{n-j} (1-a) a^n$$

$$= (1-a) a^j p^j \sum_{n=j}^{\infty} \binom{n}{j} [a(1-p)]^{n-j}$$

$$= (1-a)(ap)^j \sum_{k=0}^{\infty} \binom{j+k}{j} [a(1-p)]^k$$

$$= \frac{(1-a)(ap)^j}{[1-a(1-p)]^{j+1}} = \frac{1-a}{1-a+ap} \left(\frac{ap}{1-a+ap}\right)^j$$

where in the last step the use has been made of a relation

$$\sum_{k=0}^{\infty} \binom{j+k}{j} \alpha^k = \frac{1}{(1-\alpha)^{j+1}}.$$

Thus, $P(j) = (1-b)b^j$, with $b = ap(1-a+ap)^{-1}$; the result being the geometric distribution again, but with mean

$$ES = ap(1-a)^{-1}.$$

Note that the relation $ES = p \cdot EN$ holds again.

****

This result concerning the mean is not, however, an accident but is true for our Bernoulli trials no matter what the distribution $Q(n)$ is. Indeed, for the average number of accidents we have:

$$E(S) = \sum_j jP(j) = \sum_n \sum_j jP(j|n)Q(n) = p\sum_n nQ(n) = pE(N)$$

because for the binomial distribution

$$\sum_j jP(j|n) = np.$$

Remark: It is of great interest that our argument can be extended to more general situations. This will become clear when we shall take another look from a different angle at the problem on hand.

Consider independent identically distributed r.v.'s $Z_1, \ldots, Z_n$ with common mean $E(Z)$. Write for their sum (as in Section 5):

$$S_n = Z_1 + \ldots + Z_n.$$

This is for fixed n. When however the number of terms is a random variable itself, say N, then we get the so called random sum:

$$S = Z_1 + \ldots + Z_N$$

(also denoted by S_N, and called a compound r.v.). Then, we have from properties of conditioning:

$$pr(S=j) = \sum_n pr(S=j, N=n) = \sum_n pr(S=j \mid N=n)pr(N=n).$$

Assuming independence of N from all Z_i's, we have:

$$P(j|n) = pr(S = j \mid N = n) = pr(S_n = j).$$

Hence

$$pr(S = j) = \sum_n pr(S_n = j) pr(N = n).$$

In the special case when Z_i are Bernoulli r.v.'s (as in subsection 11.3), we obtained the situation of this section, and the formula for $pr(S = j)$ is the same as that stated at the beginning of the section.

In the general case considered now, $P(j \mid n)$ is not of the binomial form, but nevertheless we have in view of the above new look:

$$\sum_j jP(j|n) = E(S_n) = nE(Z)$$

so finally

$$E(S) = \sum_n nQ(n) = E(Z)E(N).$$

Indeed, this relation for expectation is what we would expect intuitively!

Mixing. We can extend our discussion to random sums of continuous life times with densities (as in Chapter 1). For example, a student preparing for an examination decides to solve a fixed number of problems, say n. Let $X_1, \ldots, X_n$ represent times spent on each problem, so the total time needed for all n problems is (as in Section 5)

$$S_n = X_1 + \cdots + X_n.$$

Suppose now that a student prefers rather to solve a vari-

able number N of problems, say according to mood, subject
matter, or proximity of exam date. Let distribution of N
be specified by

$$Q(n) = pr(N = n),$$

where n ranges through some set, usually n = 1, 2,... or
perhaps even n = 0, 1, 2,.... The total time for N problems
is therefore the compound life time

$$S_N = X_1 + \cdots + X_N.$$

Of great interest is the distribution of S_N and in particular its mean value.

Observe that this problem is analogous to that considered earlier in this section, but it is mixed in character. Now S_N is a continuous life time, but N is discrete.
Intuitively, we may also expect that S_N will be continuous.
We shall consider such mixed problems later in the book, so
the present discussion may serve as preparation for more
interesting applications.

To be slightly more precise, suppose that individual
life times X_i, i = 1,...,n, are i.i.d. and that N is independent of all X_i's. Then, as before, by properties of
conditioning,

$$pr(S_N \leq x) = \sum_n pr(S_N \leq x|N = n)\, Q(n) = \sum_n pr(S_n \leq x)Q(n).$$

If g_n is density of S_n (see Section 5), then denoting by h
density of S_N, one has

$$h(x) = \sum_n g_n(x)Q(n).$$

This is our answer. Unfortunately, calculations may be prohibitive, except in a few special cases. Some simplification may be achieved with the help of Laplace transforms. As in Section 7, if $f^*(\sigma)$ is the Laplace transform for individual life times, then the transform of $g_n(x)$ is

$$g_n^*(\alpha) = [f^*(\alpha)]^n,$$

so transform of $h(x)$ is given by

$$h^*(\alpha) = \sum_n [f^*(\alpha)]^n Q(n).$$

There remains of course the problem of inversion. However, useful information may be obtained without an explicit form of h. Indeed, one has immediately that, again,

$$\mathbb{E}S_N = \mathbb{E}(X) \cdot \mathbb{E}(N).$$

Let us conclude with an example. For exponential life times X_i (with parameter λ) and with a geometric distribution $Q(n) = (1 - a)^{n-1} a$ for $n = 1, 2, \ldots$, we have

$$h^*(\alpha) = \sum_{n=1}^{\infty} \left(\frac{\lambda}{\lambda + \alpha} \right)^n a(1 - a)^{n-1} = \frac{\lambda a}{\lambda a + \alpha}$$

so S_N is exponential with parameter λa.

For additional comments see Problem 31.

Chapter 2: Problems

1. At a small art exhibition there are six paintings including one by van Gogh. It is expected that on the average 2 buyers will express sufficient interest to ask for

the price of the van Gogh painting.   What is the probability of this happening?

  <u>Note</u>:   This problem is slightly ambiguous on purpose.
You may select your own interpretation.

2. A superstitious commuter driving to work must pass   n
    traffic lights, and assumes that a green light shows with
    pr.   1/2.

      The commuter believes that the day will be lucky if
    the number of green lights encountered is even, and then
    assigns the value of +1 to such a day. If the number
    of green lights is odd, the day is considered to be bad,
    and the value of   -1   is assigned to it.

      Find the average value assigned to a day (assuming
    binomial distribution.).

3. Suppose that the number of busy machines in a laundromat
    with   n   washing machines, follows the binomial distri-
    bution with   p   being the pr. that a machine is busy.

      According to the management, profit is proportional
    to the number of operating machines, except in the situ-
    ation when all   n   machines are working; the profit is
    then zero (because of high energy consumption,
    dissatisfied waiting customers, etc.).

      Thus, the profit function is

$$\varphi(j) \;=\; \begin{cases} cj & \text{for } 0 \le j \le n-1 \\ 0 & \text{for } j = n \end{cases} \;,\quad (c > 0).$$

Show that:

(i)   The average profit is

$$E\varphi(X) \;=\; np(1-p^{n-1})c.$$

(ii) The average profit has maximum when

$$p \;=\; (1/n)^{\frac{1}{n-1}}, \qquad \text{for } n > 1.$$

4. Let N be a discrete life time with binomial distribution (with parameters n and p). Suppose that the cost function φ has the form

$$\varphi(j) \;=\; z^{j} \qquad \text{for } j = 0,1,\ldots,n, \qquad |z| \le 1.$$

Show that the average cost is given by

$$E\varphi(N) \;=\; (pz + 1 - p)^{n}.$$

Sketch the graph of the average cost as a function of z, and verify that its slope at $z = 1$ is equal to the mean np.

5. Using Gaussian approximation, compute the pr. that the number of successes in 100 Bernoulli trials, with mean 80 differs in absolute value from the mean by 8 or less:

$$pr(|X-80| \le 8).$$

Compare with the Chebyshev inequality.

6. Let X be a discrete r.v. with binomial distribution
    (with parameters n and p).

    (i)   Find the second moment $\mathbb{E}(X^2)$, and plot its
          graph as a function of p (for fixed n). Verify
          that always $\mathbb{E}(X^2) \leq n^2$.

    (ii) Find p such that the second moment equals
          $k[\mathbb{E}(X)]^2$, where k is a positive integer.

    (iii) Find p which maximizes var(X), for fixed n.

7.   (i)   Sketch the graph of the binomial distribution
           P(j), for fixed j, as a function of p. Con-
           sider separately three cases: j = 0, $1 \leq j \leq n-1$,
           j = n.

     (ii) Consider for binomial distribution:

           $$F(j) \;=\; pr(X \leq j) \;=\; \sum_{i=0}^{j} P(i)$$

           defined as in the text (Section 11).
               Show that for each j

           $$F(j) \;=\; n\binom{n-1}{j} \int_{p}^{1} t^j (1-t)^{n-1-j} \, dt.$$

           Hint: differentiate both sides with respect to
           p, and use properties of binomial coefficients.

     (iii) Sketch the graph of F(j), for fixed j, as a

function of p. What are the values for p = 0,

p = 1 ?

8. Suppose that a discrete life time N has distribution
given by:

$$p(n) \quad = \quad \frac{\lambda^{n-1}}{(1+\lambda)^n} \qquad \text{for} \quad n = 1,2,\ldots$$

where $\lambda > 0$ is a constant.

(i) Show that $EN = \lambda + 1$, var $N = \lambda(\lambda+1)$.

(ii) Take $\lambda = 16/9$ and use the Gaussian approximation to
compute the probability that $N \leq 5$.

9. A repairman never shows up on the day he is called for
service, but a customer must wait for one, two or more
days. Suppose that the average waiting time is 3 days.
Find the probability of the waiting time being less than
the average (assume a geometric distribution).

10. Suppose that the average waiting time for the first success in Bernoulli trials equals $3\frac{1}{2}$.

Show that the pr. of waiting more than n (units
of time) for the first success is $(5/7)^n$.

11. A commuter driving home from work observes a "lucky"
sequence of green traffic lights (assume the geometric
distribution with p = 1/2). Find the pr. that

    (i)   The first red (i.e. not green) light occurs at the n-th traffic light intersection;

   (ii)   The first red light occurs sometime later after passing successfully n lights;

 (iii)   The first red light occurs after h additional traffic lights, having passed successfully n lights.

Compute these pr.'s for n = 1,2,3,4, and h = 1.

12. For the geometric distribution write

$$1 = \sum_{n=1}^{\infty} p(1-p)^{n-1}.$$

Differentiate both sides with respect to p and deduce that

$$\sum_{n=1}^{\infty} np(1-p)^{n-1} = 1/p.$$

13. Regard the geometric distribution

$$P(n) = p(1-p)^{n-1}$$

as a function of p, for fixed n: Show that

  (i) $dP/dp = (1-np)(1-p)^{n-2}$ for n = 1,2,... ;

 (ii) $d^2P/dp^2 = (n-1)(np-2)(1-p)^{n-3}$ for n = 1,2,...;

(iii)   P   has a maximum at $p = 1/n$ for $n = 2,3,\ldots$;

(iv)   P   has an inflection point at $p = 2/n$ for

$n = 3,4,\ldots$;

(v)   plot the graphs of   P   as a function of   p   for

$n = 1$, $n = 2$ and $n = 3$.   Indicate the slope

at $p = 0$ and $p = 1$.

14.   Let   N   be a discrete life time with geometric distri-

bution.   Suppose that the cost function φ is of the

form:

$$\varphi(n) = z^n \quad \text{where} \quad |z| \le 1, \quad n = 1,2,\ldots .$$

(i)   Show that the average cost, regarded as a func-

tion of   z,   is

$$E\varphi(N) \equiv K(z) = pz(1-qz)^{-1}.$$

(ii)   Deduce that

$$dK/dz = p(1-qz)^{-2}, \qquad d^2K/dz^2 = 2pq(1-qz)^{-3}.$$

(iii)   Verify that

$$EN = dK/dz\bigg|_{z=1} = 1/p,$$

$$EN(N-1) = d^2K/dz^2\bigg|_{z=1} = 2q/p^2$$

and deduce that

$$\text{var } N = q/p^2.$$

15. Compute for the geometric distribution:

$$pr(|N-EN| > \varepsilon), \quad \text{for} \quad \varepsilon > 0,$$

and compare it with the Chebyshev approximation.

16. Suppose that on the average 5 of your friends eat
     lunch daily at the Student Union. Using the Poisson
     distribution, calculate the pr. that on a particular
     day you will meet at lunch:

   (i)   no friend at all;

   (ii) exactly one friend;

   (iii) more than one friend.

   Do these figures agree with your experience?

17. In Criminology, it is found that the number X of
     people having finger prints belonging to a certain type
     is distributed according to the Poisson distribution
     with a parameter μ (representing the average number of
     people having the same type). For identification pur-
     poses (by detectives, in courts, etc.) it is of great
     interest to know the pr. of duplication, i.e., the pr.
     that some other person has finger prints of the same
     type as the accused.

         Formally, this is defined as the pr. of two or more
     people of the same type, when it is known that at least
     one person has this type, i.e. $\mathbb{P}(X \geq 2 \mid X \geq 1)$.

Show that the pr. of no duplication is given by

$$\mathbb{P}(X = 1 \mid X \geq 1) = \frac{\mu}{e^{\mu}-1} \, .$$

Calculate this pr. for $\mu = 1/2$, $\mu = 1$, $\mu = 5$. Does this pr. decrease with increasing μ ? Try to sketch the graph of this pr. as a function of μ.

18. Suppose that the pr. of an individual having an infection is $p = 1/25$. Compute the pr. that in a group of 100 people more than 6 people will have an infection, using the

(i) Poisson approximation;

(ii) Gaussian approximation.

19. If there are on the average 1% left handers, find the pr. of having at least 5 left handers amoung 300 people.

20. Estimate the number of raisins which a cookie should contain on the average if it is desired that the pr. of a cookie to contain at least one raisin to be 0.99 or more.

21. Let X be a discrete r.v. with Poisson distribution (with parameter μ).

(i)   Show that

$$\text{var}(X/\sqrt{\mu}) \ = \ 1.$$

(ii)   Let $Y = (X - \mu)^2$. Show that

$$EY^2 \ \geq \ \mu^2.$$

22.   Regard the Poisson distribution

$$P(j), \qquad j = 0,1,\ldots, \quad \mu > 0$$

as a function of its mean μ, for fixed j. Show that:

(i) $dP/d\mu = P(j-1)(1-\mu/j)$ for $j \neq 0$, and

$dP/d\mu = -P(0)$ for $j = 0$;

(ii) $d^2P/d\mu^2 = P(j-2)[1 - \frac{2j\mu-\mu^2}{j(j-1)}]$ for $j = 2,3,\ldots$

$\qquad\qquad = \ P(0)(\mu-2)$ for $j = 1$

$\qquad\qquad = \ P(0)$ for $j = 0$.

(iii) P has maximum for $j = 1,2,\ldots$ at $\mu = j$

(iv) P has two inflection points at $\mu = j \pm \sqrt{j}$, for
$j = 2,3,\ldots$, and one inflection point at $\mu = 2$
for $j = 1$.

(v)   Plot graphs of P as a function of μ for
$j = 0,1,2$ and 3.

23.   Show that for the Poisson distribution $F(j) = \sum_{i=0}^{j} P(i)$
can be written as

$$F(j) = \frac{1}{j!} \int_{\mu}^{\infty} t^j e^{-t} \, dt$$

(see Section 14). Show that $dF(j)/d\mu = -P(j)$.

24. Write the Erlang formula for blocking in the form

$$B = P(s)/F(s)$$

(see Example 6, Section 14), and deduce that

$$dB/d\mu = B[B + (s-\mu)/\mu].$$

25. Let N be a discrete life time with Poisson distribution. Suppose that the cost function φ is of the form:

$$\varphi(j) = z^j \quad \text{where} \quad |z| \le 1, \quad j = 0,1,\dots \ .$$

   (i) Show that the average cost, regarded as a function of z, is

$$E\phi(N) = K(z) = e^{-\mu(1-z)}.$$

   (ii) Deduce that

$$dK/dz = \mu K(z), \qquad d^2K/dz^2 = \mu^2 K(z).$$

   (iii) Verify that

$$EN = dK/dz\Big|_{z=1} = \mu,$$

$$EN(N-1) \quad = \quad d^2K/dz^2 \Big|_{z=1} \quad = \quad \mu^2$$

and deduce that

$$\text{var } N = \mu.$$

26. Suppose that the life time N has Poisson distribution
     with mean μ. Let the cost function be $(-1)^n$ for
     $n = 0,1,\ldots$. Show that the average cost is $e^{-2\mu}$.
     (Note: this is immediate! Contrast it with Example 3
     in Section 14).

27. Compute for the Poisson distribution

$$pr(|N-EN| > \varepsilon), \quad \text{for} \quad \varepsilon > 0,$$

and compare it with the Chebyshev approximation.

28. Let X be a discrete life time assuming values
     $0,1,2,\ldots,K$ (finite or infinite). Suppose that the
     cost function φ is of the form:

$$\varphi(j) = z^j \quad \text{for} \quad |z| \leq 1.$$

Then, the average cost, regarded as a function of z,
and defined by

$$C(z) \quad = \quad E(z^X)$$

has properties analogous to the Laplace transforms (see
Section 7):

(i) $C(z)\Big|_{z=1} = 1$, $dC(z)/dz\Big|_{z=1} = E(X)$,

$d^2C(z)/dz^2\Big|_{z=1} = EX(X-1)$.

(ii) Let $X_1,\ldots,X_n$ be independent identically distributed life times with common distribution with average cost $C(z)$. Consider the total life time $S_n = X_1 + \ldots + X_n$. Show that the average cost for S_n is

$$E(z^{S_n}) = [C(z)]^n$$

(this is the analogue of convolutions in Section 5).

(iii) Show that if X assumes values 0 and 1 only, with pr.'s q and p, respectively, then S_n has the binomial distribution with parameters n, p. (This is the same derivation as that in subsection 11.3, see also Problem 4.)

(iv) Show that if $X_1,\ldots,X_n$ have common binomial distribution with parameters k, p, then S_n has also a binomial distribution but with parameters nk, p, (see Problem 4).

(v)   Show that if $X_1,\ldots,X_n$ have common Poisson distribution with mean μ, then S_n has also Poisson distribution but with mean $n\mu$ (see Problem 25).

(vi)   Show that if $X_1, \ldots, X_n$ have common geometric distribution pq^{j-1} $(j = 1, 2, \ldots)$, then S_n has so called "negative binomial distribution" of the form

$$P(j) \;=\; \binom{j-1}{j-n} p^n q^{j-n}, \quad \text{for} \quad j \geq n$$

(see Problem 14).

29. With reference to discrete life times $Z_1, \ldots, Z_n$ and S and N in Section 15, consider the following average costs:

$$C(z) = E(z^Z), \qquad U(z) = E(z^N) = \sum_{n=0}^{\infty} z^n Q(n),$$

$$W(z) \;=\; E(z^S) \;=\; \sum_{j=0}^{\infty} z^j P(j).$$

(i)   Using the formula for $P(j)$, show that

$$W(z) \;=\; U[C(z)].$$

(ii) Deduce that

$$E(S) \;=\; E(Z)E(N)$$

$$\text{var } S \;=\; (\text{var } Z)E(N) + (\text{var } N)(EZ)^2.$$

(iii) Using the above relations, check var S in the examples worked out in Section 15.

30. Suppose that in Problem 29: $C(z) = 1 - \sqrt{1-z}$ and that

$Q(n)$ is the Poisson distribution with mean λ. Show that

$$W(z) = e^{-\lambda\sqrt{1-z}} .$$

What about $E(S)$?

31. With reference to the Laplace transform $h^*(\sigma)$ from Section 15 (mixing), show that in the case of i.i.d. life times with the common transform $f^*(\alpha)$,

$$h^*(\alpha) = C(f^*(\alpha))$$

where $C(z)$ is the cost function defined in Problem 28 as

$$C(z) = \sum_n Q(n)z^n.$$

(i) Verify that

$$- \log h^*(\alpha) = m\sigma/(\lambda + \alpha)$$

for Poisson $Q(n)$ with mean m and for exponential lifes (with mean $1/\lambda$).

(ii) Verify that:

$$h^*(\alpha) = q + pg_n^*(\alpha)$$

when Q has the form

$$Q(0) = q, \qquad Q(n) = p \qquad \text{for a fixed n,}$$
$$q + p = 1.$$

32. Let X be the number of successes in n Bernoulli trials.

(a)   Suppose that $pr(X = 0) = pr(X = 1)$. Find

$$pr(X > 0),   pr(X = j|X > 0)$$

for $0 < j \leq n$.

(b)   Suppose that $pr(X = 0) = pr(X = n)$. Find

$$pr(X \geq 1),   pr(X = j|X \geq 1)$$

for $1 \leq j \leq n$.

33.   Suppose that in Bernoulli trials the average number of successes in n trials is m.

(i)   Consider the cost function ϕ defined by

$$\phi(j) = \begin{cases} (n - j)/n & \text{for } j = 1, 2,\ldots,n. \\ 0 & \text{for } j = 0. \end{cases}$$

Show that the average cost associated with the number X of successes is

$$\mathbb{E}\phi(X) = (1 - m/n) - (1-m/n)^n.$$

Explain what happens when $n = 1$.

(ii) Let N be the waiting time for the first success. Show that the pr. of waiting more than n for the first success is

$$(1 - m/n)^n.$$

What happens when $n \to \infty$?

34.   Suppose that the cost function ϕ in n Bernoulli trials with pr. of success p has the form

$$\phi(0) = \phi(n) = c > 0,   \text{and}$$

$$\phi(j) = 0   \text{otherwise,}$$

where c is a constant.

(i)   Calculate the average cost as a function of p, and show that it has a minimum when p = 1/2 for n > 1. What, when n = 1?

(ii)   Plot the graph of the average cost as a function of p.

35. Let P(j), j = 0, 1,...,n, be a binomial distribution with parameters n and p. Let Q(k), k = 0, 1,...,n, be another binomial distribution with parameters n and q, where p + q = 1. Define

$$R(j) = pP(j) + qQ(n - j) \qquad \text{for } j = 0, 1,...,n.$$

Show that R = P.

36. Consider two binomial distributions with the same n but with different p. With the first distribution, pr. of no success in n trials is the same as pr. of no failure in n trials. With the second distribution, the mean is three times larger than variance.

(i)   Find the mean of each distribution.

(ii)   Assuming the common cost function given by

$$\phi(j) = (n - j)c, \quad 0 \le j \le n, \; c > 0, \text{ calculate}$$

the average cost for each distribution.

37. In a workshop with n machines, the average number of working machines is m, where m is an integer such that 0 < m < n.

Suppose that the cost of operation is proportional to the number of working machines, except when exactly m

machines are busy in which case the cost is zero.
Thus, the cost function is

$$\phi(j) = \begin{cases} cj & \text{for } j \neq m, \\ 0 & \text{for } j = m, \end{cases}$$

where c is a positive constant.

Find the average cost of operation, and show that it
is smaller than cm.

38. Consider Bernoulli trials with a success representing
breakage of a machine.

   (i)   Suppose that the waiting time for a machine to
         break down is m days, on average. Find the pr.
         that the waiting time is exactly m days.

   (ii) Suppose now that there are n such machines in a
         workshop. Find the pr. that exactly one of
         them is broken.

   (iii) Calculate the above pr.'s when n = 3 and m = 3
         days.

   (Does the problem make sense if this m is expressed as
   a fraction of a month, say m = 1/10 month?)

39. On a given day in a workshop, the number of machines to
be used may be either one with pr. a > 0 or two with
pr. b > 0, with a + b = 1.

   Each of these machines may fail with pr. p. Let S
be the number of machines that failed ("accidents"),
and let P(j) be its distribution (j = 0, 1, 2). Find
P(j) and the mean value of S.

40. Mother of a young Peter, pleased with his decision to
     collect money for some worthy cause instead of candy
     during a recent Halloween night, said to him, "For each
     house that contributes to your collection I will give
     you \$2 from my purse and an extra \$2 if none of the
     houses you visit gives you anything." Assuming that
     Peter visited a few houses in the neighborhood and that
     each house offered him money with the same pr. p, find
     the average gain Peter may expect to get from his
     mother. Note that Peter's gain function is

$$\phi(j) = 2j \qquad \text{for } j \neq 0 \qquad \text{and}$$
$$\phi(0) = 2 \qquad \text{for } j = 0.$$

Show that the average gain is an increasing function of
p.

41. Let P(j) be the pr. of j successes in n Bernoulli
     trials. Suppose that the pr. p of a success is given
     by a d.f. F of some life time:

$$p = F(x) \qquad \text{for fixed } x, \text{ with } 0 \leq x < \infty.$$

Assume that for all $x > 0$, both $F(x)$ and its density
$f(x)$ are positive.

(i) For fixed n and fixed j, consider the ratio

$$H(x) = P(j)/P(j + 1)$$

regarded as a function of x for $0 < x < \infty$, and
$j = 0, 1, \ldots, n - 1$. Show that H is a decreasing
function of x. What are the limiting values of
H when $x \to 0$ and $x \to \infty$?

(ii) Suppose that n is even. Let S(x) be the pr. that half the trials result in successes. Find the expression for S(x), and show that S(x) — treated as a function of x—reaches maximum for x such that

$$F(x) = 1/2.$$

42. Suppose that in some learning experiment it takes on average m trials to obtain the correct response for the first time. Let N be the waiting time for the first correct response. Assuming the geometric distribution,

   (i) Show that:

$$pr(N > m) = \left(\frac{m - 1}{m}\right)^m.$$

   (ii) Show that:

$$pr(|N - m| > m) \le (m - 1)/m.$$

43. Suppose that a discrete life time has the Poisson distribution with mean μ. Denote by $C(\mu)$ the average cost as a function of μ, and show that:

   (i) $C(\mu) = b(\mu + 1)^2$ when the cost function is $\phi(j) = b(j^2 + j + 1)$, where b is a constant.

   (ii) $C(\mu) = -e^{-\mu} + \mu - 1$ when the cost function is

$$\phi(j) = \begin{cases} j - 1 & \text{for } j = 1, 2, \ldots \\ -2 & \text{for } j = 0 \end{cases}$$

   and that $C(\mu)$ increases with μ.

44. Let X and Y be two i.i.d. r.v.'s having the Poisson distribution with mean μ. Show that

$$pr(X = i, Y = j \mid X + Y = i + j) = \binom{i + j}{i} (1/2)^{i+j}$$

for i, j = 0, 1, 2,... .

45. Suppose that the joint distribution of a continuous life time X, assuming values in the unit interval (0,1), and a discrete life time Y, assuming values 0, 1,...,n, is specified by the following expression:

$$pr(X \le x, Y = k) = \int_0^x \binom{n}{k} t^k (1 - t)^{n-k} \, dt$$

for $0 \le x \le 1$, $0 \le k \le n$

(with obvious specification outside the indicated region). Show that $\mathbb{E}X = 1/2$, $\mathbb{E}Y = n/2$, and $\mathbb{E}(XY) = n/3$.

3
To Renew or Not to Renew

Renewal processes are very intuitive and rather easy to handle (at least at the beginning--in later stages things start to be awkward), and have many practical applications. Indeed, we shall talk now about situations familiar from everyday life -- replacement or renewal of worn out appliances, devices etc., and other successive life times.

Although the mathematical aspects may be more complex than what we have seen so far, you should have no difficulty in following the presentation, if treated with compassion. Renewals are described in Section 16, and Section 17 presents the glorious achievements of the theory. One may often wonder why so much work is needed to justify so "obvious" results. As a consolation, we may be happy to see that our mathematics agrees with our intuition.

The story of the rabbit (Section 18) as sad as it may be, should stop us for awhile to reflect on what lies in the background. We only remark here that we touched upon the extension of renewals to random walks, and on the modern

concept of stopping times (a vast treasure for the mathematically minded!).

Section 16:   Renewals

Consider the following homely situation. A light bulb
is shining brightly in your room, and suddenly it blows up.
You replace it by another one -- it lasts for a duration of
its life and again burns itself out. You replace it again,
and again it blows up, so again... . Two obvious questions
present themselves:

1)   What is the distribution of the total time needed to use
     all your supply of bulbs?

2)   What is the distribution of the total number of bulbs
     during a year?

In the first question the r.v. is the total life time, in
the second question the r.v. is the number of bulbs; the
first one is the continuous r.v., whereas the second is discrete. We have been discussing such r.v.'s separately, and
now it is time to consider them jointly. It is clear that
both questions are related to each other. It may be more
convenient for the reason of generality to speak about
renewals, that is immediate replacements of objects (bulbs,
cars, machines, persons, etc.) at the instant of termination
of a life time. Thus we would like to have a distribution
of

1)   total life time for n renewals,

2)   total number of renewals within time t.

Denote by X_n the life time of the object immediately pre-
ceeding the n-th renewal (life time of the n-th bulb),
where n = 1,2,···· . We shall assume that all life times
X_n are independent, identically distributed (i.i.d.) r.v.'s
with a common d.f. F having density f, and with mean   μ
and variance σ^2. That is we are in the situation in Section 2:

$$pr(X_n \leq t) = F(t) \quad \text{for} \quad 0 \leq t < \infty \quad \text{(independent of n).}$$

Total time for n renewals is clearly

$$S_n = X_1 + \ldots + X_n.$$

Write for its d.f.:

$$pr(S_n \leq t) = G_n(t), \qquad 0 \leq t < \infty, \quad n = 1,2,\ldots .$$

Clearly, S_n are nonnegative, and G_1 = F. We already
have noted that $E(S_n) = n\mu$, and (by independence) $var(S_n)$
= $n\sigma^2$; see Section 5.

Our first problem will be solved if we can find the
distribution of S_n. We have done it already in Section 5.
Consider first n = 2, so $S_2 = X_1 + X_2$. As is shown in Sec-

tion 5, S_2 has density g_2 given by the convolution

$$g_2(t) = \int_0^t f(t-s)f(s)\, ds.$$

Now, write $S_n = S_{n-1} + X_n$; r.v.'s S_{n-1} and X_n are independent, and we can apply to their densities the same argument, to get

$$g_n(t) = \int_0^t f(t-s)g_{n-1}(s)\, ds, \quad n = 2,3,\ldots, \quad 0 \le t < \infty.$$

Thus, starting with $g_1 = f$, we may compute all g_n from the above recurrence relation. Unfortunately, actual calculations are usually complicated.

It may be convenient to define $S_0 = 0$, so its d.f. $G_0 = 1$ for all t. Then, remembering that G_n is the integral of g_n (for $n = 1,2,\ldots$), we can write the above recurrence relation in the form

$$\boxed{G_n(t) = \int_0^t F(t-s)\, dG_{n-1}(s), \quad 0 \le t < \infty, \quad n = 1,2,\ldots}.$$

Recall that G_n is the pr. that the total life time of n renewals is t or less. Moreover, it has been assumed that the 0-th renewal (i.e. the beginning of the first life time) is at the origin. This provides the solution for question 1.

The collection of r.v.'s (S_n) is called a <u>renewal</u> <u>process</u>.

Remark: X_n represents a life time of the n-th item (installed in a certain system), that is the time interval between the (n-1)-th and the n-th renewals. S_n is then the total time for n renewals, or the time at which the n-th renewal takes place.

Although the G_n are hard to compute, they are helpful to solve Problem 2. Recall that we wish to find the distribution of a number of renewals in a fixed time interval from 0 to t (excluding the instant 0, but including the instant t). More precisely, define for each t a r.v. N_t representing the number of renewals that will have occurred by the time t, including any made at t but excluding the initial (the 0-th). Write for the distribution of N_t:

$$pr(N_t = n) = P_t(n), \quad n = 0,1,\ldots, \quad t > 0.$$

Observe that $P_t(n)$ is the distribution in n; it represents the number of renewals that may take place. In particular, for n = 0, $P_t(0)$ is the pr. that no renewal took place; that is the original item is still in progress at time t. Clearly, the r.v. N_t is discrete. It should be stressed that t is a parameter. Thus, we have a family of r.v.'s (N_t) -- such a family is known as a stochastic process -- (N_t) is also called a renewal process. And for each t, the r.v. N_t has its distribution P_t.

To find P_t, note that the value of N_t gives the index n for which the next sum S_{n+1} overshoots the point

t for the first time. Hence, the event of exactly n
renewals up to the time t is:

$$(N_t = n) \quad = \quad (S_n \le t < S_{n+1}) \quad = \quad (S_n \le t) - (S_{n+1} \le t).$$

Taking pr.'s of both sides, one finds that

$$\boxed{P_t(n) \quad = \quad G_n(t) - G_{n+1}(t), \quad n = 0,1,2,\ldots, \quad t > 0} \, .$$

This is the basic relation, solving Problem 2. Note that
for n = 0 we have obviously:

$$P_t(0) = 1 - F(t).$$

It is of interest to note that in view of the convolution
expression for G_{n+1} stated above, we can write:

$$P_t(n) \quad = \quad \int_0^t [1-F(t-s)] \, dG_n(s), \quad n = 1,2,\ldots \, .$$

This relation is useful on many occasions.

Another relation between P_t and G_n is obtained as
follows. Write $P_t(n)$ for each n:

$$P_t(0) \quad = \quad 1 \quad\quad - G_1(t)$$

$$P_t(1) \quad = \quad G_1(t) \quad - G_2(t)$$

$$\vdots \quad\quad\quad\quad \vdots \quad\quad\quad\quad \vdots$$

$$P_t(n-1) \quad = \quad G_{n-1}(t) - G_n(t).$$

Adding, one has

$$\sum_{k=0}^{n-1} P_t(k) \ = \ 1 - G_n(t), \qquad n = 1,2,\ldots$$

which is nothing else but

$$\mathrm{pr}(N_t \le n-1) \ = \ \mathrm{pr}(S_n > t)$$

a useful relation which expresses duality between N_t and S_n.

It can be verified that

$$G_n(t) \ = \ \sum_{k=n}^{\infty} P_t(k), \qquad n = 0,1,2,\ldots \ .$$

In particular, for $n = 0$, this shows that for each $t > 0$:

$$\sum_{k=0}^{\infty} P_t(k) \ = \ 1.$$

It is clear from this that $\displaystyle\lim_{n\to\infty} G_n(t) = G_\infty(t) = 0$ identically, so S_n increases to ∞ with pr. one.

Example: Suppose that the common lifetime is exponential

$$F(t) \ = \ 1 - e^{-\lambda t}, \quad t \ge 0, \quad \lambda > 0,$$

with density

$$f(t) \ = \ \lambda e^{-\lambda t}, \qquad t > 0, \quad \lambda > 0.$$

As shown in Section 5, $S_2 = X_1 + X_2$ has density

$$g_2(t) = \lambda t e^{-\lambda t}, \qquad t > 0.$$

Try once more (for $n = 3$):

$$g_3(t) = \int_0^t \lambda s e^{-\lambda s} \lambda \cdot \lambda e^{-\lambda(t-s)} \, ds = \frac{(\lambda t)^2}{2} e^{-\lambda t} \lambda.$$

It can be verified (by induction) that the g_n have the general form:

$$g_n(t) = \frac{(\lambda t)^{n-1}}{(n-1)!} e^{-\lambda t} \lambda, \qquad\qquad t > 0.$$

Hence, the d.f. G_n is

$$G_n(t) = \int_0^t \frac{(\lambda \tau)^{n-1}}{(n-1)!} e^{-\lambda \tau} \lambda \, d\tau$$

(by substitution $\lambda \tau = x$)

$$= \frac{1}{(n-1)!} \int_0^{\lambda t} x^{n-1} e^{-x} \, dx$$

Integrating by parts, or using the integral from Section 14):

$$= 1 - \sum_{k=0}^{n-1} \frac{(\lambda t)^k}{k!} e^{-\lambda t}.$$

Hence,

$$P_t(n) = G_n(t) - G_{n+1}(t) = \sum_{k=0}^{n} \frac{(\lambda t)^k}{k!} e^{-\lambda t} - \sum_{k=0}^{n-1} \frac{(\lambda t)^k}{k!} e^{-\lambda t}$$

$$= \frac{(\lambda t)^n}{n!} e^{-\lambda t}, \qquad n = 0,1,2,\ldots, \qquad t > 0.$$

Thus, we get the Poisson distribution!! This derivation of the Poisson distribution is much more important than that from Section 14. It stresses the importance of the Poisson distribution on its own merits.

Note: $P_t(0) = e^{-\lambda t}$.

Section 17: Renewal Equation

In the previous section we found the distribution P_t of the number N_t of renewals up to time t (including t). We also have seen that the calculations may be cumbersome. Thus, it will be of more interest to find the <u>average</u> <u>number</u> <u>of</u> <u>renewals</u> <u>up</u> <u>to</u> t, that is $E(N_t)$. It is a pleasant surprise that this expected value can be calculated directly, without explicit knowledge of the distribution P_t.

Write for the <u>renewal</u> <u>function</u>:

$$U(t) = E(N_t)$$

Then using results from the previous section:

$$U(t) = \sum_{n=0}^{\infty} nP_t(n) = \sum_{n=1}^{\infty} n[G_n(t) - G_{n+1}(t)]$$

$$= G_1(t) - G_2(t) + 2[G_2(t) - G_3(t)] + 3[G_3(t) - G_4(t)] + \ldots$$

$$= G_1(t) + G_2(t) + G_3(t) + \ldots$$

$$= \sum_{n=1}^{\infty} G_n(t).$$

This is a very interesting formula; the drawback is that we need all G_n. However, we can do much better than that! Proceed as follows:

$$U(t) = G_1(t) + \sum_{n=2}^{\infty} G_n(t) \qquad \text{splitting the sum}$$

$$= G_1(t) + \sum_{n=2}^{\infty} \int_0^t F(t-s)dG_{n-1}(s) \qquad \begin{array}{l}\text{using recurrence}\\ \text{relation for } G_n\end{array}$$

$$= G_1(t) + \int_0^t F(t-s) \, d \sum_{n=2}^{\infty} G_{n-1}(s) \qquad \text{change of order}$$

$$= G_1(t) + \int_0^t F(t-s) \, dU(s) \qquad \text{definition of } U(s)$$

Recall that $G_1 = F$, so we have the <u>renewal equation</u> for $U(t)$:

$$\boxed{U(t) = F(t) + \int_0^t F(t-s) \, dU(s)} \, .$$

This is the famous renewal equation -- it allows us to find $U(t)$ using only the known d.f. F -- thus avoiding finding individual G_n.

Define the <u>renewal density</u> $u(t) = dU(t)/dt$. Remember that $U(t)$ is the expected number of renewals up to time t (and it is <u>not</u> the d.f.), so $u(t)$ may be regarded as the expected number of renewals per unit of time; more precisely, the average number of renewals during the interval from t to t + h is

$$U(t+h) - U(t) = \int_t^{t+h} u(s)\ ds.$$

Differentiating the renewal equation one obtains the renewal equation for the density u:

$$u(t) = f(t) + \int_0^t f(t-s)u(s)\ ds, \qquad t > 0.$$

Thus, knowledge of the density f of life times is sufficient to determine density u of renewals. This is a rather remarkable result. It has many applications in various fields.

Since the equation for u involves u also under the integral sign, the renewal equation has the form of an integral equation. Special techniques are needed for finding a solution. However, in the present case a solution can be found by a very simple method, namely the Laplace transform defined in Section 7.

Laplace transforms of densities f and u are defined by the following integrals:

$$f^*(\alpha) = \int_0^\infty e^{-\alpha t} f(t)\ dt, \qquad u^*(\alpha) = \int_0^\infty e^{-\alpha t} u(t)\ dt, \qquad \alpha \geq 0.$$

Write

$$w(t) = \int_0^t f(t-s)u(s)\ ds.$$

Simple evaluation of the double integral shows that the
Laplace transform of w(t)   is

$$w^*(\alpha) \;=\; \int_0^\infty e^{-\alpha t} w(t)\; dt \;=\; f^*(\alpha) u^*(\alpha)$$

(This is expressed by saying that the Laplace transform of a
convolution is a product of Laplace transforms.)

   Now taking Laplace transforms of both sides of the
renewal equation for density, one finds that

$$u^*(\alpha) \;=\; f^*(\alpha) + f^*(\alpha) u^*(\alpha).$$

Hence

$$\boxed{\; u^*(\alpha) \;=\; \frac{f^*(\alpha)}{1 - f^*(\alpha)} \;}.$$

Thus, to find the renewal density, we first compute the
Laplace transform of the density f,   then calculate the
Laplace transform u^* from the above formula. It then
remains to invert $u^*(\alpha)$ to find u(t). This last step can
be facilitated by the use of tables of Laplace transforms.

   Thus, we may consider our problem of finding the renewal
function U(t)   as solved.

   It may be proved that as time increases the renewal
density u tends to a constant value, equal to the reciprocal of the average life time $\mu = E(X)$. This is the famous
renewal theorem:

$$\boxed{\lim_{t\to\infty} u(t) \;=\; 1/\mu}\;.$$

Indeed, if this limit exists, then it can be found from the following property of Laplace transform (mentioned in Section 7):

$$\lim_{t\to\infty} u(t) = \lim_{\alpha\to 0} \alpha u^{*}(\alpha)$$

Using the expression for $u^{*}(\alpha)$ obtained earlier, one finds with the help of the L'Hôpital rule that indeed the value of the limit is $1/\mu$ (see Problem 21).

Thus, whatever the life time distribution, the average number of renewals up to t is approximately (for large t):

$$\boxed{\Pi(t) \;\approx\; t/\mu}$$

and clearly tends to infinity as $t \to \infty$. This agrees with the intuitive interpretation.

In another formulation, the average number of renewals in the interval $(t,t+h)$ for $h > 0$, is approximately for large t:

$$\boxed{U(t+h) - U(t) \;\approx\; h/\mu}$$

for any form of life time distribution with mean life time μ. The quantity $1/\mu$ is called a rate of renewals, because it represents a number of renewals per unit of time (for large t).

$$**** ************ ****$$

Poisson Example: The case of exponential life time and the associated Poisson distribution (discussed in the previous section) is of great importance in applications.

We have here

$$f^*(\alpha) \quad = \quad \int_0^\infty e^{-\alpha t} \lambda e^{-\lambda t} \, dt.$$

Hence

$$u^*(\alpha) \quad = \quad \lambda/\alpha$$

and this implies that $u(t) = \lambda$ is a constant. Hence the limit is of course λ, too, and one has exactly:

$$U(t) = \lambda t.$$

In other words, $E(N_t) = \lambda t$, as it could be seen from the fact that N_t has Poisson distribution.

To avoid confusion, let us repeat that the mean number of renewals is λt, whereas $1/\lambda = \mu$ is the mean life time (between renewals).

Example 2: Suppose that customers arriving to a shop (or telephone calls at an exchange, insurance claims, etc.) can be described by a renewal process. That is, let X_n be the time interval between the arrival of the $(n-1)$-th and n-th customer. Suppose that the average interarrival time is $\mu = 5$ min.

Then, approximately the average number of arrivals within one hour is $t/\mu = 60/5 = 12$ (irrespective of the form of distribution of interarrival times).

Assume now that interarrival times have exponential density. Then, the pr. of having exactly 4 customers within a quarter of an hour is

$$P_{15}(4) = 3^4 e^{-3}/4! = 0.168$$

because $\lambda t = 15/5 = 3$.

Example 3: Suppose that the average number of cars passing during a (12-hour) day is 360. Find the average number of cars during 2 hours. Here we have approximately

$$U(t+2) - U(t) \approx 2/\mu = 2(320/12) = 60$$

because $\mu = 12/360$. What is the pr. of having exactly 60 cars in a 2 hour period, assuming a Poisson distribution? This can be read off from the tables, but in this range our tables are insufficient. One can use the Gaussian approximation, as in Section 9:

$$\text{pr}(59 < N_t \le 60) = \text{pr}(-1/8 < Z \le 0) = \Phi(1/8) - 1/2$$

$$= 0.550 - 0.500 = 0.050,$$

recalling that the standard deviation is $\sqrt{60} \approx 8$.

Example 4: Suppose that the life time has uniform distribution with density $1/L$ over the interval $(0,L)$. Then

$$f^*(a) = (1-e^{-aL})/aL$$

and

$$u^*(a) = (1-e^{-aL})/(e^{-aL}-1+aL).$$

It can be verified that

$$\lim_{a \to 0} au^*(a) = \frac{2}{L}$$

as it should. Hence, for $h > 0$

$$U(t+h) - U(t) \approx 2h/L.$$

Counterexample. As noted, given f determines u through the renewal equation. The converse need not be true, however. Indeed, the selection of arbitrary renewal density u, even if $u(t) \to 1/\mu$ as $t \to \infty$, does not imply in general that the inverted renewal equation

$$f^*(a) = \frac{u^*(a)}{1 + u^*(a)}$$

should give a density. Indeed, suppose that

$$u(t) = (1-e^{-\mu t})/\mu \to 1/\mu \qquad \text{as } t \to \infty.$$

The Laplace transform is

$$u^*(a) = \frac{1}{\mu}\left(\frac{1}{a} - \frac{1}{a+\mu}\right)$$

and therefore

$$f^*(a) = 1/(a^2 + \mu a + 1).$$

Suppose now that $\mu < 2$. Then, it can be verified by integration that the function

$$f(t) = e^{-\mu t/2}\frac{\sin(\zeta t/2)}{(\zeta/2)}, \quad 0 < t < \infty,$$

where $\zeta = \sqrt{4-\mu^2}$, has the above $f^*(a)$ as its Laplace transform. But of course, this $f(t)$ is not a density (why?), although $\int_0^\infty f(t)\, dt = 1$.

Example 6: Let g be the density of the waiting time in the bus problem (Section 4) associated with the density f of the interarrival time. Let us take this waiting time as our life time X of the renewal process. It is easy to see that the Laplace transform of g is

$$g^*(a) = \frac{1 - f^*(a)}{a\mu}$$

where μ is the mean of f. Then,

$$u^*(a) = \frac{1 - f^*(a)}{a\mu - 1 + f^*(a)}.$$

The renewal theorem yields:

$$\lim_{\alpha \to 0} \alpha u^*(\alpha) \ = \ \lim_{t \to \infty} u(t) \ = \ 1/w$$

where w is the average waiting time found in Section 4.
As a matter of fact we can use this observation to derive
the expression for w.

Extensions. S_n is the total time for n renewals. Replacing
fixed n by a random variable N_t, one has a new random variable S_{N_t} which represents the total time needed for the random number of renewals.

   Thus, S_{N_t+1} - t is the time interval from an arbitrary instant t (following a renewal) until the next renewal. That is precisely the remaining life time since t.
This is the situation discussed in connection with a bus
problem; the distribution of the waiting time for the next
renewal we found in Section 4 corresponds to the distribution of S_{N_t+1} - t, as t goes to infinity (so called equilibrium conditions.

   We can verify this fact by deriving the explicit
form of the distribution of the waiting time $W_t = S_{N_t+1}$ - t.
The following derivation makes interesting use of the renewal equation and needs the rather intuitive notion of
conditioning (as explained in Section 3):

$$\text{pr}(W_t \leq x) = \text{pr}(S_{N_t+1} \leq x + t) = \sum_n \text{pr}(S_{n+1} \leq x + t, N_t = n)$$

$$= \sum_n \text{pr}(S_n \leq t < S_{n+1} \leq t + x)$$

$$= \sum_n \int_0^t d \, \text{pr}(S_n \leq s) \text{pr}(t < S_{n+1} \leq t + x | S_n = s)$$

$$= \sum_n \int_0^t dG_n(s) \, [F(x + t - s) - F(t - s)]$$

$$= F(x + t) - F(t) + \int_0^t [F(x + t - s)$$

$$- F(t - s)] \, dU(s)$$

$$= F(x + t) - \int_0^t F^C(x + t - s) \, dU(s)$$

$$= F(x + t) - \int_x^{x+t} F^C(s) u(x + t - s) \, ds$$

$$\rightarrow 1 - \int_x^\infty F^C(s) \, ds/\mu \qquad \text{as } t \rightarrow \infty,$$

which is the distribution from Section 4.

Note that the dual life time $A_t = t - S_{N_t}$ represents the age of an item at instant t, that is, the time interval from the last renewal before an arbitrary time t. It is of interest to remark that although A_t has a different distribution from that of W_t, nevertheless both W_t and A_t have the same limiting distribution.

Consider now a "random interval" straddling a fixed time instant t, namely,

$$L_t = S_{N_t+1} - S_{N_t} = A_t + W_t.$$

Intuitively, one would expect that its distribution is the same as that of the renewal life times, namely F. Well, it is not. Indeed, note that in equilibrium the mean of L_t is different from μ:

$$\text{or} \quad \mathbb{E} \, L_t = \mu(1 + \frac{\sigma^2}{\mu^2}) \geq \mu.$$

This is of course another aspect of the pecularities of the bus problem.

Section 18: Jumping Rabbit

A rabbit runs across field and suddenly falls into a deep ditch. To get out, the rabbit jumps several times until the magnitude of a jump is larger than the height of the ditch. What is the pr. that the rabbit will be free on the n-th attempt? What is the average number of attempts? We tacitly assume that the rabbit is determined and does not tire easily.

Many situations can be described by the jumping rabbit model. Suppose that there is a certain critical level (like a noise threshold, the level of water in a river, the boiling temperature, the height of a ditch). Observations of a life time are made and and long as the observed values stay below the critical level, no action is taken. At the instant when the critical level is reached, (for the first time), an alarm is sounded. Of interest is the number of attempts for the first occurrence of the alarm.

18.1.

Denote by X_n the magnitude of the n-th attempt (say, the n-th jump), $n = 1,2,\ldots$, and let x be the critical level (height of a ditch). As long as $X_n \leq x$ attempts will continue, until the first overshot over x ($X_n = x$ means the rabbit's nose reached the surface; to get out the rabbit needs more than x .

Let N be a r.v. representing the successful attempt of the first jump over the level x. Clearly:

$$N = \min(n: n \geq 1, X_n > x).$$

Write for the distribution of N:

$$pr(N = n) = P_x(n), \qquad n = 1,2,\ldots ,$$

where x is a parameter denoting the critical level.

To make the problem precise, further assumptions must be imposed. The following ones are the simplest (although not always too realistic):

A1:   attempts are made independently of each other; that
      is, r.v.'s X_n are independent.

A2:   the magnitude of each attempt is governed by the same
      distribution with density f and mean μ; that is,
      all life times X_n have the same distribution.

The indicated procedure implies that the event "rabbit
is out at the n-th attempt" is:

$(N = n)$ $=$ $(X_1 \leq x, X_2 \leq x, \ldots, X_{n-1} \leq x, X_n > x)$ for $n = 2, 3, \ldots$

$(N = 1)$ $=$ $(X_1 > x)$ for $n = 1$.

Hence, by independence:

$$\mathrm{pr}(N = n) = \mathrm{pr}(X_1 \leq x, \ldots, X_{n-1} \leq x, X_n > x)$$

$$= \mathrm{pr}(X_1 \leq x) \ldots \mathrm{pr}(X_{n-1} \leq x) \mathrm{pr}(X_n > x)$$

$$= F(x) \ldots F(x) F^c(x) = F^{n-1}(x) F^c(x), \quad \text{for } n = 1, 2, \ldots$$

In other words we get the geometric distribution:

$$\boxed{P_X(n) = pq^{n-1}, \qquad n = 1, 2, \ldots}$$,

where

$$p = F^c(x), \qquad q = F(x).$$

From Section 13 we know that the average number of attempts
will be:

$$E(N) = 1/F^c(x)$$

and the pr. of some more than n attempts is $pr(N > n) = F^n(x)$.

Example: (i) For the exponential distribution of rabbit's jumps:

$$p = e^{-\lambda x}, \qquad q = 1 - e^{-\lambda x}$$

so

$$P_x(n) = (1-e^{-\lambda x})^{n-1}e^{-\lambda x}, \qquad E(N) = e^{\lambda x}, \qquad pr(N > n) = (1-e^{-\lambda x})^n$$

ii) For the uniform distribution of rabbit's jumps:

$$p = 1 - x/L, \qquad q = x/L$$

where L is the maximum height of a jump, so

$$P_x(n) = (1-x/L)(x/L)^{n-1}, \qquad E(N) = L/(L-x), \qquad pr(N > n) = (x/L)^n$$

where $0 < x < L$; clearly if $x \geq L$, then $p = 0$ and the rabbit will never get out.

18.2.

Consider now a different version of a jumping rabbit. A rabbit starts from an initial position, say after a nap, jumps here and there executing a "random walk" with jumps of various magnitude. What is the pr. that a rabbit will fall in a ditch?

To simplify the discussion, we shall assume that jumps occur along a straight line. Let X_n be the magnitude of the n-th jump, $n = 1,2,\ldots$. Positive values of X_n mean that a jump to the right occurred, and negative values of X_n indicate a jump to the left. Assume that (X_n) are i.i.d. with common d.f. F having density f and mean μ. Suppose that the initial position of the rabbit is at the origin, so $S_0 = 0$, and let S_n denote the rabbit's position after n jumps:

$$S_n = X_1 + \ldots + X_n, \qquad n = 1,2,\ldots .$$

Observe that although (X_n) are independent, (S_n) are dependent r.v.'s.

Select a fixed positive t, and take the interval (t,∞) as the "taboo set" representing a ditch. If $S_n \le t$ the rabbit enjoys free jumps; as soon as $S_n > t$ for the first time, the rabbit fell into a ditch and the process stops. If N is a r.v. representing the number of jumps to reach the ditch, then clearly:

$$p(N = n) = \text{pr}(S_1 \le t, \ldots, S_{n-1} \le t, S_n > t).$$

Explicit determination of this pr. is tough, but several general conclusions may be drawn easily. For convenience, consider the event $A_n = (S_n > t)$, and note that its complement is $A_n^c = (S_n \le t)$. Hence we can write

$$\text{pr}(N = n) = \text{pr}(A_1^c \ldots A_{n-1}^c A_n).$$

Consider now the disjoint representation of a union of sets:
$\cup_{i=1}^{n} A_i = \cup_{i=1}^{n} A_1^c \dots A_{i-1}^c A_i$. Hence

$$pr(\cup_{i=1}^{n} A_i) = \sum_{i=1}^{n} pr(A_1^c \dots A_{i-1}^c A_i) = pr(N \leq n)$$

or

$$pr(N > n) = pr(\cap_{i=1}^{n} A_i^c) = pr(S_1 \leq t, \dots, S_n \leq t) \equiv G_n(t).$$

Thus, $G_n(t)$ denotes the pr. that the rabbit did not over-shoot t in the first n jumps, or equivalently that falling into a ditch will occur after more than n jumps.

It is clear now that

$$pr(N = n) = G_{n-1}(t) - G_n(t), \quad \text{for } n = 1, 2, \dots$$

with $G_0(t) = 1$. Alternatively, this relation follows from the observation that:

$$pr(A_1^c \dots A_{n-1}^c A_n) = pr(A_1^c \dots A_{n-1}^c) - pr(A_1^c \dots A_n^c).$$

Hence, by the same argument as in Section 17, the average number of jumps for the rabbit is:

$$E(N) = \sum_{n=1}^{\infty} n \, pr(N = n) = \sum_{n=0}^{\infty} G_n(t) = 1 + \sum_{n=1}^{\infty} G_n(t).$$

And evidently, S_N is the position at which the poor rabbit landed in the ditch.

Recalling the definition of $M_n = \max(S_1, \ldots, S_n)$ from Section 6, one can see that

$$G_n(t) = pr(M_n \le t).$$

It should be stressed that G_n is the d.f. of M_n, and not of S_n (that was the case of renewals discussed in Section 16 because of positivity of life times). Indeed, the pr. of not crossing the barrier at t is the same as the pr. that the maximum of positions did not cross t.

However, it can be deduced that (see Problem 17)

$$G_n(t) = \int_{-\infty}^{t} G_{n-1}(t-s) \, dF(s), \qquad n = 1, 2, \ldots$$

with $G_0 = 1$. In principle, the above recurrence relations could be used for evaluation, but calculations are prohibitive, and special methods are required.

Letting $n \to \infty$, one obtains:

$$G_\infty(t) = pr(N = \infty) = pr(S_n \le t \text{ for all } n)$$

This pr. is zero for most situations. If it is not zero, then it will satisfy the integral equation

$$G_\infty(t) = \int_{-\infty}^{t} G_\infty(t-s) \, dF(s).$$

The task of finding a solution of this integral equation is fascinating, but regretfully we cannot discuss it here (the transform method used earlier does not work here!). We shall limit ourselves to several comments which are rather intuitive, although their mathematical justification is very delicate.

It can be shown that if $\mu > 0$, the random walk of our rabbit will drift to $+\infty$ (visiting the positive half-axis infinitely often) and if $\mu < 0$, then the random walk will drift to $-\infty$ (visiting the negative half-axis infinitely often). When $\mu = 0$, then the random walk will oscillate, visiting positive and negative half-axes infinitely often.

Thus, for $\mu > 0$ the rabbit is bound to fall into the ditch (jump over t), so N will be finite, and this means that $G_\infty(t) = 0$ identically; in addition $E(N) < \infty$. As a matter of fact, one even has $E(S_N) = \mu E(N)$ as in Section 15.

On the other hand, for $\mu < 0$ the rabbit may fall into the ditch (having made some jumps to the right), or may not. Indeed, our rabbit may be lucky and never even approach

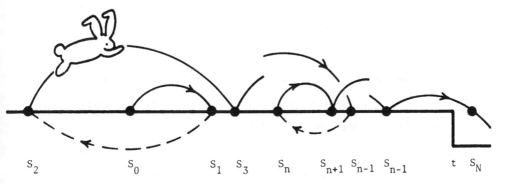

Fig 18.1
Jumping rabbit.

the ditch (stays below t). This means that N will assume
infinite value with positive probability. Thus, $G_\infty(t)$ =
$pr(N = \infty) > 0$; hence also $E(N) = \infty$. As a matter of fact,
one has $G_\infty(\infty) = 1$, so G_∞ is the proper d.f.

Finally when $\mu = 0$, the hesitant rabbit will wander
along the field, and will eventually fall in. Thus, $G_\infty(t)$
= 0 again, so N is finite but then $E(N) = \infty$. It will
take awfully long on the average to reach the ditch if one
really does not care where one is going.

Chapter 3: Problems

1. Suppose that interarrival times between consecutive buses
    (arriving at a bus stop) form a renewal process; that is
    let X_n be a time interval between the (n-1)-th and
    n-th bus. Assume that the X_n are exponential life
    times with mean $\mu = 15$ minutes.

    Find the pr. of having two or more buses within
    a time interval of length t. Call this pr. b(t), and
    plot its graph as a function of t.

2. Mary has an electric gadget which works on a single
    battery. As soon as the battery in use fails, Mary
    immediately replaces it with a new battery.

    If the life time of a battery is (in hours) uni-
    formly distributed over the interval (20,40), then at
    what rate does Mary have to change batteries?

How many batteries on the average will Mary use per month (assume here that Mary uses the gadget only 5 hours per day)?

3. Peter buys records at the rate of 2 per month.

   (i)   How many records will he buy during a year?

  (ii)   Assuming the Poisson distribution, what is the pr. that during the summer (i.e., 2 1/2 months) he will buy

      (a)   exactly 3 records,

      (b)   more than 3 records,

      (c)   at least one record?

4. Let $P_t(n)$ be the Poisson distribution with mean λt.

   (i)   Show that:

$$\int_0^\infty P_t(n) e^{-qt}\, dt = \lambda^n / (\lambda+q)^{n+1}$$

where $q \geq 0$. For $q = 0$, the integral gives $1/\lambda$ for all n.

  (ii)   Evaluate:

$$\int_0^\infty t^m P_t(n)\, dt = \frac{(n+m)!}{n!} \frac{1}{\lambda^{m+1}} .$$

 (iii)   Plot the graph of the function $P_n(t)$ for $t > 0$, for fixed $n = 0, 1, \ldots$. Indicate the shape, find the maximum, inflection points, etc. (Consider separately $n = 0$ and $n \neq 0$).

5. Let $P_t(n)$ be the Poisson distribution with mean λt.
    Write $Q_n(t) = dP_t(n)/dt$. Show that

    (i) $Q_n(t) = \lambda P_t(n-1) - \lambda P_t(n)$ for $n \geq 1$, and
    $$= -\lambda P_t(0) \qquad \text{for } n = 0.$$

    (ii) $$\int_0^\infty Q_n(t)\, dt = 0 \quad \text{for } n \geq 1,$$
    and $$= -1 \quad \text{for } n = 0.$$

6. Show that for each k:
    $$\lim_{t\to\infty} \sum_{n=k+1}^{\infty} \frac{(\lambda t)^n}{n!} e^{-\lambda t} = 1.$$

7. Suppose that the r.v. N_t has the Poisson distribution
    with mean λt. Let the cost function be $(-1)^n$, $n =$
    $0,1,2,\ldots$. Show that the average cost is
    $$\sum_{n=0}^{\infty} (-1)^n \frac{(\lambda t)^n}{n!} e^{-\lambda t} = e^{-2\lambda t}.$$

    Note: No calculation needed! Compare this with
    Example 3 in Section 14).

8. Let f be the density of a life time of the following
    form:
    $$f(x) = \lambda \frac{(\lambda x)^{n-1}}{(n-1)!} e^{-\lambda x}, \qquad n = 1,2,\ldots, \qquad x \geq 0$$
    Show that
    $$\int_0^t f(x) e^{-\lambda(t-x)}\, dx$$
    gives the Poisson distribution with mean λt.

9. Suppose that the life time in a renewal process has uni-
    form distribution over the time interval (a,a+L) where
    a and L are positive constants.

    Show that the average number of renewals during the
    interval (t,t+h), with h > 0, is given approximately
    (for large t) by 2h/(2a+L).

10. Suppose that the life time in a renewal process has the
    d.f. of the form:

    $$F(t) = \frac{t^2}{(1+t^2)} \qquad \text{for } t \geq 0.$$

    Show that the average number of renewals during the
    interval (t,t+h), with h > 0, is given approximately
    (for large t) by: $2h/\pi$.

11. Suppose that the life time in a renewal process has d.f.
    of the following form:

    $$F(x) = 1 - e^{-\lambda\sqrt{x}} \qquad \text{for } x \geq 0$$

    where $\lambda > 0$ is a constant.

    Show that the average number of renewals during the
    time interval (t,t+h), with h > 0, is given approxi-
    mately (for large t) by $\lambda^2 h/2$.

12. Suppose that the life time in a renewal process has
    density of the form

    $$f(t) = cae^{-at} + (1-c)be^{-bt}, \qquad \text{for } t \geq 0$$

    where $a \neq b$, $a > 0$, $b > 0$ and $0 < c < 1$.

Show that the average number of renewals during
the interval (t,t+h), with h > 0, is given approximately (for large t) by:

$$abh/(a+cb-ca).$$

13. Suppose that the life time in a renewal process has
density of the form

$$f(t) = \begin{cases} c\lambda e^{-\lambda t} + (1-c)\lambda, & \text{for } 0 < t < 1/\lambda \\ c\lambda e^{-\lambda t}, & \text{for } 1/\lambda < t < \infty, \end{cases}$$

where $\lambda > 0$ and $0 < c < 1$.

Show that the average number of renewals during
the interval (t,t+h), with h > 0, is given approximately (for large t) by:

$$2\lambda h/(1+c).$$

14. Suppose that the density f of the common life time X
in the renewal process is supported by the interval
[s,s+L], as in Section 2. Then, clearly the total
life times S_n take values in the interval [ns,ns+nL].
Show that:

$$\text{pr}(N_t \geq n) = \begin{cases} 0 & \text{for } t < ns \\ 1 & \text{for } t \geq n(s+L). \end{cases}$$

15. Jack keeps a gold fish in a fish bowl of height L,
and checks the level of water each morning. If the
level exceeds x inches (where 0 < x < L), he does

nothing. If the level drops below x, he fills the bowl to capacity with water.

Suppose that the distribution of the water level is uniform on the interval (0,L). Show that the probability that the bowl is filled on the n-th day for the first time is:

$$p(n) \;=\; (x/L)(1 - x/L)^{n-1}, \qquad n = 1,2,\ldots$$

and that the average number of days until the first fill is L/x.

16. In the jumping rabbit problem, let $P_x(n)$ be the pr. that the rabbit escapes at the n-th attempt, assuming exponential distribution of the rabbit jumps.

Show that for fixed $n \geq 2$, this pr. has the maximum value for the ditch depth x equal to:

$$x \;=\; \lambda^{-1} \log n.$$

What for n = 1 ?

17. Proceeding as in Section 5, show that $G_2(t)$ from Section 18.2 can be expressed as:

$$G_2(t) \;=\; pr(S_1 \leq t, \; S_2 \leq t) \;=\; \iint_{\substack{x \leq t \\ x+y \leq t}} f(x)f(y) \; dx \; dy$$

$$=\; \int_{-\infty}^{t} F(t-s)f(s) \; ds.$$

Then, proceeding by induction, derive the recurrence relation for $G_n(t)$ stated in Section 18.2.

18. Take $f(x) = \frac{1}{2} e^{-|x|}$ and using Problem 17 show that

$$G_2(t) = \begin{cases} (3/8) e^t & \text{for } t \le 0 \\ 1 - (5/8) e^{-t} - (1/4) t e^{-t} & \text{for } t \ge 0 \end{cases}$$

and $E(M_2) = 1/2$.

Try for $G_3(t)$, if you have the patience.
Note, however, that now $\mu = 0$, so necessarily
$G_\infty(t) = 0$ for all t.

19. Suppose that jumps X_n in Section 18.2 have density of
the form:

$$f(t) = \begin{cases} \frac{\lambda\mu}{\lambda+\mu} e^{\lambda t}, & \text{for } t < 0 \\ \frac{\lambda\mu}{\lambda+\mu} e^{-\mu t}, & \text{for } t > 0 \end{cases}$$

where $\lambda > 0$ and $\mu > 0$ are constant and $\rho = \lambda/\mu < 1$.
Hence $E(X_n) = (\rho-1)/\lambda < 0$ (actually, X_n is the
difference of two exponential life times).

Verify that the d.f. G_∞ defined by:

$$G_\infty(t) = \begin{cases} (1-\rho) e^{\lambda t}, & \text{for } t \le 0 \\ 1 - \rho e^{-\mu(1-\rho)t}, & \text{for } t \ge 0 \end{cases}$$

satisfies the integral equation

$$G_\infty(t) = \int_{-\infty}^{t} G_\infty(t-s) f(s) \, ds$$

for all t.

Verify that G_∞ has a density which is a continuous function of t.

20.   In the renewal process, let $u^*(\alpha)$ be the Laplace trans-
      form of the renewal density $u(t)$. Using $u^*(\alpha)$, show
      that for the renewal function $U(t)$ one has in the limit

$$\lim_{t \to \infty} U(t) = \infty.$$

21.   With reference to renewals in Section 17, show that

$$\lim_{\alpha \to 0} \alpha u^*(\alpha) = 1/\mu.$$

22.   Suppose that the life time has uniform distribution
      with density $1/L$ over the interval (O,L), as in Example
      4 in Section 17 (see also Problem 14).

      (a)   Show that for $t \geq O$ (in notation of Section 17:

$$G_n(t) = \frac{1}{n!L^n} \sum_v (-1)^v \binom{n}{v} (t - Lv)^n,$$

      where v ranges over $O \leq vL \leq t$.

      (b)   Also show that:

$$U(t) = \sum_{v=0}^{n} \left(\frac{t - Lv}{L}\right)^v \frac{(-1)^v}{v!} e^{-(t-Lv)/L}$$

      where n is determined by the relation $nL \leq t \leq$
      $(n + 1)L$.

23.   Suppose that the common life time density in renewal
      process has the form

$$f(t) = \frac{(\lambda t)^{n-1}}{(n - 1)!} \lambda e^{-\lambda t}, \qquad t > O,$$

where the integer n and λ are fixed constants. Show
that the renewal density is

$$u(t) = \sum_{r=1}^{\infty} \frac{(\lambda t)^{nr-1}}{(nr - 1)!} \lambda e^{-\lambda t}$$

and

$$\lim_{t \to \infty} u(t) = \lambda/n.$$

24. Suppose that the life time of renewals has the distri-
     bution function F with density

$$f(t) = (2\lambda)^2 t e^{-2\lambda t}, \qquad t \geq 0.$$

  (a)   Show that the renewal function has the form
$$U(t) = \lambda t - (1 - e^{-4\lambda t})/4, \qquad t \geq 0$$
         with renewal density
$$u(t) = \lambda(1 - e^{-4\lambda t}).$$

  (b)   Show that the number of renewals in a period
         $(t, t + h)$ for large t is on average approxi-
         mately λh.

  (c)   Show that the distribution of the waiting time
         W_t (given at the end of Section 17) is

$$pr(W_t \leq x) = 1 - e^{-2\lambda x} - \lambda x e^{-2\lambda x} (1 + e^{-4\lambda t}),$$
$$t \geq 0, \quad x \geq 0.$$

25. In the renewal process, let N_t be the number of
     renewals up to time t, and let F be the life time
     distribution. Denote by G_n the n-fold convolution of

F, with $G_1 = F$.   Let $U(t)$ be the renewal function. Define

$$Z(t) = \mathbb{E} N_t^2 + U(t).$$

(i)   Show that

$$Z(t) = \sum_{n=1}^{\infty} 2nG_n(t)$$

(ii)   Deduce that $Z(t)$ satisfies the integral equation

$$Z(t) = 2U(t) + \int_0^t Z(t - s) \, dF(s).$$

(iii)   Let $z(t) = dZ(t)/dt$, and denote by $z^*(\alpha)$ the Laplace transform of $z(t)$.   Show that the solution of the above integral equation may be written in the transformed form as

$$z^*(\alpha) = 2u^*(\alpha)/(1 - f^*(\alpha)), \qquad \alpha > 0,$$

where $u^*(\alpha)$ and $f^*(\alpha)$ are Laplace transforms of densities $u(t)$ and $f(t)$, respectively.

(iv)   Calculate $Z(t)$ for exponential renewals.

26.   In the renewal process, let $P_t(n)$ be the pr. of exactly n renewals up to time t.   Denote its derivative by

$$q_t(n) = dP_t(n)/dt.$$

(i)   Show that:

$$q_t(n) = g_n(t) - g_{n+1}(t) \qquad n = 1, 2, \ldots ,$$

$$q_t(0) = -f(t) \qquad n = 0,$$

where $g_n(t)$ is the density of the total life S_n and $f(t) = g_1(t)$.

(ii)   Write $q_\alpha^*(n)$ for the Laplace transform of $q_t(n)$. Show that:

$$q_\alpha^*(n) = [1 - f^*(\alpha)] \, [f^*(\alpha)]^n, \qquad n = 1, 2, \ldots,$$

$$q_\alpha^*(0) = -f^*(\alpha),$$

where $f^*(\alpha)$ is the Laplace transform of $f(t)$. Verify that

$$\sum_{n=0}^{\infty} q_\alpha^*(n) = 0.$$

27.   Let $H(t)$ be a distribution function of some life time, with density $h(t)$, Laplace transform $h^*(\alpha)$, and mean m.   The average number of renewals during a random time interval whose length has distribution $H(t)$ is given by

$$A(t) = \int_0^\infty (U(t + s) - U(t))/dH(s), \qquad t \geq 0.$$

(i)   Show that:

$$A(t) = \int_0^t u(t - s) \, H^c(s) \, ds,$$

where $u(t)$ is the renewal density.

(ii)   Show that:

$$\int_0^\infty A(t) e^{-\alpha t} dt = u^*(\alpha)(1 - h^*(\alpha))/\alpha,$$

$$\alpha > 0.$$

(iii)   Show that:

$$\lim_{t \to \infty} A(t) = \lambda m,$$

where $u(t) \to \lambda$.

(iv)   Show that when $H(t)$ coincides with the d.f. of

renewals $F(t)$, then $A(t) = F(t)$.

28.   Suppose that the life time distribution in the renewal

process has the form

$$F(t) = 1 - \rho e^{-(1-\rho)\mu t}, \qquad t \geq 0,$$

where $0 < \rho < 1$ and $\mu \geq 0$ are constants. Note that

$F(0) = 1 - \rho > 0$, so there is no density f, in viola-

tion of our assumption in Section 16. This distribu-

tion occurs in Queueing problems—see Problem 15 in

Chapter 4, where its transform $f^*(\alpha)$ has been found

(note that μ is here a parameter, not the mean value).

Substituting this transform into equation for $u^*(\alpha)$,

show that:

$$u^*(\alpha) = (1 - \rho)(\alpha + \mu)/(\alpha\rho), \qquad \alpha > 0.$$

Deduce then that

$$\lim_{t \to \infty} u(t) = \mu(1 - \rho)/\rho, \qquad U(0) = (1 - \rho)/\rho > 0.$$

29.   Suppose that the life time distribution in the renewal

process has the form

$$F(t) = \gamma(1 - e^{-\lambda t}), \qquad t \geq 0,$$

where $0 < \gamma < 1$ and $\lambda > 0$ are constants. Note that

$F(\infty) = \gamma$, so the life time may be infinite with pr.
1 - γ (in violation of our standing assumption on life
times).

Nevertheless, this distribution has a density f(t),
which of course integrates only up to $\gamma < 1$.

Show that the Laplace transform of f(t) and of the
renewal density u(t) are

$$f^{*}(\alpha) = \gamma\lambda/(\alpha + \lambda),$$

$$u^{*}(\alpha) = \gamma\lambda/[\alpha + \lambda(1 - \gamma)], \qquad \alpha > 0,$$

and deduce that

$$\lim_{t\to\infty} u(t) = 0$$

and

$$U(t) = (1 - e^{-\lambda(1-\gamma)t}) \; \gamma/(1 - \gamma) \to \gamma/(1 - \gamma) < \infty$$

$$\text{as } t \to \infty.$$

30. Suppose that a life time X has density f of the form

$$f(x) = \begin{cases} 0 & \text{for } 0 < x < a, \\ \lambda e^{-\lambda(x-a)} & \text{for } a < x < \infty, \end{cases}$$

where $\lambda < 0$ and $a > 0$ are constants. Show that:

(i)   The average life time is

$$\mathbb{E}\,X = a + 1/\lambda.$$

(ii) In the renewal process, with this f as a life
      time density, the average number of renewals
      in the period (t, t + h) for large t is approxi-
      mately $\lambda h/(1 + \lambda a)$.

31. Suppose that in some industrial problem the hazard rate
     r has the form

$$r(t) = t \geq 0.$$

   (i)   Compute the hazard function R.

  (ii) Find the d.f. F, determined by R, for the life
     time X.

 (iii) Find the mean life time $\mathbb{E}X$.

  (iv) Suppose now that X is the life time in a re-
     newal process. Show that the average number of
     renewals during the interval (t,t + h), with
     h > 0, is given approximately (for large t) by
     $h\sqrt{2/\pi}$.

32. Let X and Y be two independent life times with exponen-
     tial distribution, but with different parameters λ and
     μ, respectively. Let

$$Z = X + Y.$$

   (i)   Find density of Z and $\mathbb{E}Z$.

  (ii) Suppose that Z is the life time in a renewal
     process. Show that the average number of
     renewals during the interval (t,t + h), with
     h > 0, is given approximately (for large t) by
     $\lambda\mu h/(\lambda + \mu)$.

 (iii) Show that the renewal density is

$$u(t) = [1 - e^{-(\lambda+\mu)t}]\lambda\mu/(\lambda + \mu).$$

  (iv) Regard u(t) as a cost function associated with
     life time X. Show that the average cost is

$$\mathbb{E}u(X) = \int_0^\infty f(t)u(t)dt$$

$$= 2\lambda\mu(\lambda + \mu)(2\lambda + \mu)^{-1}(2\mu + \lambda)^{-1}.$$

33. In the problem of a rabbit trying to escape from a
     ditch of height x > 0, the pr. of escape at the n-th
     attempt has been found in Section 18.1 to be

$$p_x(n) = F^c(x)(F(x))^{n-1}$$

for an arbitrary distribution function F.

(i)   For fixed n >1, find the height x_0 for which
      this pr. is maximum.

(ii) Show that with this x_0, the average number of
      jumps will be n.

(iii) Consider special cases of uniform, exponential
      distributions (e.g., Problem 16). Plot graphs
      of $p_x(n)$ as a function of x for fixed n.

34. In the Poisson renewals consider distribution $P_t(n)$
     with mean λt and n = 0, 1, 2,... .

(a)   Define the following probability:

$$b(t) = P_t(n) + P_t(n + 1), \qquad n \geq 1.$$

Considering b(t) as a function of t (for fixed n),
show that:

(i)   There is a maximum at t_0, given by $\lambda t_0 = \sqrt{n(n + 1)}$.

(ii) Slope at t = 0 equals λ for n = 1 and 0 for n >
      1.

(iii) Total area under the curve b($\cdot$) equals $2/\lambda$.
      Sketch the graph of b(t).

(b)   Define the following probability:

$$b(t) = P_t(0) + P_t(1).$$

Considering b(t) as a function of t, show that:

(i)   b(t) is strictly decreasing.

(ii) b(t) has the inflection point at the value of
      t equal to the mean life time.

(iii) The total area under the curve b($\cdot$) equals $2/\lambda$.
      Sketch the graph of b(t).

(c)   Let X be a life time with density f defined,
      for fixed n $\geq$ 0, by

$$f(t) = \lambda b(t)/2, \qquad t \geq 0.$$

Show that in both cases

$$\mathbb{E}\,X = (2n + 3)/(2\lambda), \qquad \text{for each } n \geq 0.$$

4

Markovian Dance

In many practical situations we are interested in random fluctuations of a number of elements or components of a certain kind. Typically, we may investigate the number of people in the waiting line, or the number of customers in a shop, or the number of animals in a certain region. The discrete life time which expresses this number depends itself on time, so its probability distribution will also be a function of time. In other words, we have a family of discrete r.v.'s, say Y_t, where t is time, and this family $(Y_t, 0 \leq t < \infty)$ is called a stochastic process (with discrete state space and continuous parameter).

There is a class of stochastic processes which is characterized by some kind of lack of memory, in the sense that the future development of the process depends only on the present, but not the past. There are many important practical problems which have (or may be thought to have) this property. You may recall that our old friend the exponential life time has it. Such processes are known as Markov processes, and we shall study in this Chapter many interest-

ing problems in which fluctuations dance to a Markovian tune.
Actually, we shall consider examples which belong to a nar-
rower category, appropriately called the "birth and death"
process. Indeed, we regard increase in the number as births
and decrease as deaths (think about a colony of bacteria).
These processes have been investigated extensively in recent
years both from a theoretical point of view and for practical applications.

     In Section 19 we shall examine the Poissonian input and
get another derivation of the Poisson distribution. Section
20 treats the learning model and Section 21 a group of lines;
both sections hint at another point of view (related to
renewals). Sections 22 and 23 treat the simple queueing
model of great importance in applications. Section 24 deals
with general births, exclusively. The method of analysis is
that of difference-differential equations for transition
pr.'s. These equations are derived individually for each
case. This seems to be very instructive. However, in Section 25 the general theory of birth and death processes is
illustrated from the modern point of view, based on the
"balance equation" (Chapman-Kolmogorov), and the reader is
probably left baffled by all that stuff. We could probably
begin this chapter with Section 25 and derive others from it.
But in this way we may lose a lot of charm peculiar to the
individual special cases. So it is better to leave Section
25 at the end, and content ourselves with a brief indication
of what is in store by alluding to strange things in the

preliminary Section 24. The reader who wishes to investigate matters further is strongly encouraged to look at Section 25. Finally, in the Section 26 we consider branching models of great importance in population studies.

Again, this chapter leans heavily on the previous chapters, but its mathematics is perhaps a little more difficult. And, unfortunately, calculations become lengthy, too. But that is the nature of our subject, and we may as well start to like it.

Section 19: Poisson Input

Consider a "system" (a shop, a news stand, an office, a factory, a telephone exchange, etc.) to which "customers" arrive (people, machines, letters, messages, claims, calls, etc.). We shall be interested in the number of customers who arrived within a given time interval -- i.e., the input to the system. Actually, we shall consider a most popular kind of input, known as the Poisson input.

Denote by Y_t a number of customers present in the system at time t. Here $0 \leq t$, and Y_0 is the number of customers present initially. Clearly, $Y_t - Y_0 = N_t$ is the number of those who arrived during the time interval $(0,t)$. Similarly, the number of arrivals during time interval $(t,t+h)$ is $Y_{t+h} - Y_t$. Poisson input is specified by the following assumptions:

A1: The pr. that during an interval $(t,t+h)$ where $h > 0$ is small, an arrival occurs is approximately equal to λh.

A2: The pr. of more than one arrival during (t,t+h) is
     negligible.

A3: Arrivals in disjoint time intervals are independent
     ("independent increments").

Physically, this implies that the total number of incoming
customers can only increase in unit steps, with the pr. of
a new customer independent of time and of the number of
customers present, and proportional to the interval length
h. The factor of proportionality λ is called the intensity,
or <u>arrival</u> <u>rate</u>.

   We shall be interested in the conditional pr. that at
time t there are j customers in the system, when at time
0 there have been i customers. In symbols:

$$p_{ij}(t) \;=\; \mathrm{pr}(Y_t = j \mid Y_0 = i), \qquad 0 \le t, \quad i,j = 0,1,2,\ldots \; .$$

The event $(Y_t = j)$ is frequently expressed by saying that
at t the system is in state j. Thus, $p_{ij}(t)$ expresses
the <u>transition</u> <u>pr</u>. from state i (initially) to state j
at time t. Since the number of customers can only increase,
it is clear that

$$p_{ij}(t) = 0 \qquad \text{for } i > j.$$

   Assumptions A1-A3 are used to derive equations for the
transition pr. For this purpose, consider two contiguous
intervals (0,t) and (t,t+h), where h is small. Suppose
that at t = 0 there are initially i customers, and at
time t+h there are j. Clearly j ≥ i. If at least one
arrival occurred during the time interval (0,t+h), then

one must distinguish the three mutually exclusive ways for
this to happen:

(a) no arrival during (t,t+h) and j-i arrivals during
     (0,t), i.e., transition from i to j;

(b) one arrival during (t,t+h) and j-i-1 arrivals
     during (0,t), i.e., transition from i to j-1;

(c) two or more arrivals during (t,t+h) and less than
     j-i-1 arrivals during (0,t), i.e., transition from
     i to less than j-1.

In terms of pr.'s the above contingencies may be expressed
by the "balance equation":

$$p_{ij}(t+h) = p_{ij}(t)p_{jj}(h) + p_{ij-1}(t)p_{j-1,j}(h) + \cdots$$

The pr. of the third contingency being negligible by A2.
Now, by A1 one has approximately

$$p_{j-1,j}(h) = \lambda h.$$

Hence, the pr. of no arrival during (t,t+h) is:
$p_{jj}(h) = 1 - \lambda h$, approximately. Consequently, the balance
equation becomes (up to approximation):

$$p_{ij}(t+h) = p_{ij}(t)(1-\lambda h) + p_{ij-1}(t)\lambda h + \cdots, \quad \text{for small } h.$$

Hence,

$$\frac{p_{ij}(t+h) - p_{ij}(t)}{h} = -\lambda p_{ij}(t) + \lambda p_{ij-1}(t) + \cdots .$$

Passing to the limit as $h \to 0$, one obtains a difference-differential equation:

$$\frac{dp_{ij}(t)}{dt} = -\lambda p_{ij}(t) + \lambda p_{ij-1}(t), \qquad \text{for } j > i \qquad (*)$$

For $j = i$, only the first contingency can occur, so by the similar reasoning

$$P_{ii}(t+h) = p_{ii}(t)(1-\lambda h) + \ldots$$

which leads to:

$$\frac{dp_{ii}(t)}{dt} = -\lambda p_{ii}(t), \qquad \text{for } j = i. \qquad (**)$$

The initial condition is expressed by:

$$P_{ii}(0) = 1, \qquad p_{ij}(0) = 0 \quad \text{for } j \neq i. \qquad (***)$$

There are several methods for solving this infinite system of difference-differential equations. The most natural one, although not the simplest, is the step-by-step method. Starting with the differential equation (**) and using the initial condition (***), it is easily seen that

$$p_{ii}(t) = e^{-\lambda t}, \qquad \text{for } t \geq 0.$$

Remark: Observe that the above expression is in fact the pr. of no arrival up to t, that is of no change in state

i. In other words, this is the complementary d.f. of the exponential life time (the interarrival period). This is in agreement with the discussion of the Poisson process in terms of renewals. Furthermore, one has for small h:

$$e^{-\lambda h} = 1 - \lambda h + \ldots$$

indicating that A1 indeed corresponds to the pr. of termination being approximately λh. See Section 3.

The next step is the solution for $j = i + 1$. One has here the linear differential equation

$$\frac{dp_{i,i+1}(t)}{dt} + \lambda p_{i,i+1}(t) = \lambda e^{-\lambda t}, \qquad \text{with} \quad p_{i,i+1}(0) = 0.$$

Then, for $j = i + 2$, and so on. It is, however, much simpler to observe that differentiation of the Poisson distribution with respect to t, carried out in Section 17, yields equations analogous to (*) and (**). Thus, one can easily verify by differentiation that the solution is:

$$p_{ij}(t) = \frac{(\lambda t)^{j-i}}{(j-i)!} e^{-\lambda t}, \qquad \text{for } j \geq i, \quad t \geq 0.$$

Thus, we have again obtained the Poisson distribution; hence the name-Poisson input. Note that this is actuallly the third derivation of the Poisson distribution; in fact this is the most common one.

What we have just shown indicates that the Poisson input is simultaneously an example of a Markov process with

independent increments and also a counting (renewal) process with exponential inter-arrival times.

****

Suppose that we are given the initial distribution of
Y_0: $\text{pr}(Y_0 = i) = \pi_i$. Then the joint distribution of Y_0
and Y_t is (by rules of conditional pr.):

$$\text{pr}(Y_0 = i, \ Y_t = j) \ = \ \pi_i P_{ij}(t).$$

Hence, the distribution of $N_t = Y_t - Y_0$, the number of
arrivals during $(0,t)$ is:

$$\text{pr}(N_t = n) \ = \ \sum_{i=0}^{\infty} \pi_i P_{i,i+n}(t) \ = \ P_t(n)$$

where $P_t(n)$ is the Poisson distribution as in Section 16,
in agreement with the definition of N_t.

****

The <u>conditional</u> <u>expectation</u> of Y_t given that $Y_0 = i$,
is defined by

$$E(Y_t \mid Y_0 = i) \ = \ \sum_{j=0}^{\infty} j P_{ij}(t).$$

This is the average number of customers present at time t,
given that initially there were i customers. Easy computation shows that the above conditional expectation is:

$$\sum_{j=i}^{\infty} [(j-i)+i] P_{ij}(t) \ = \ \sum_{n=0}^{\infty} n P_t(n) + i \sum_{j=i}^{\infty} P_{ij}(t) \ = \ \lambda t + i$$

which is intuitively obvious, as $E(N_t) = \lambda t$ is the expected number of arrivals during $(0,t)$ and i is the initial number. Note that the conditional expectation is not a number, but a function of i.

Example: If the number of accidents since the beginning of this week has been 4, and the rate is 7 accidents per week, what is the pr. of an accident tomorrow? Here $i = 4$, $j = 5$ and $\lambda = 7/7 = 1$ per day, and $t = 1$ day. So $\lambda t = 1$, $n = j-i = 1$, and $P_1(1) = 0.3679$.

The average number of accidents since the beginning of the week until tomorrow will be $i + \lambda t = 4 + 1 = 5$.

****

Random time:

Consider again the Poisson input, and assume for convenience that initially (at $t = 0$) there were no arrivals $(i = 0)$. Thus, N_t is the number of arrivals during $(0,t)$, with $N_0 = 0$, and the distribution of N_t is Poissonian with mean λt, namely $p_{0j}(t)$ for $j = 0,1,\ldots$.

Consider now, instead of an interval of fixed (deterministic) length t, an interval of random length T. Here T is a life time of an interval. We would like to find the distribution of arrivals during such random intervals. We shall write N_T in analogy to N_t, to stress the fact that duration of the interval is now T. We shall assume that T is independent of the Poisson input.

This situation is very common in applications. For

example, in the telephone system T may represent the "hold-ing time" of a call (or duration of a conversation), and it is required to find the distribution of incoming calls dur-ing a holding time. In other application, T may be dura-tion of the observation interval (for road traffic, say) and N_T gives the number of arrivals during that period.

As N_t is a well behaved (discrete) life time, and T a life time well familiar to us, we may rest assured that N_T will also be an honest (discrete) life time. Indeed, when $T = t$, then N_T becomes N_t. Hence, we can regard distribution of N_t as that of N_T, but conditional on $T = t$:

$$pr(N_T = j \mid T = t) \quad = \quad \frac{(\lambda t)^j}{j!} \, e^{-\lambda t} \quad = \quad pr(N_t = j).$$

Let f be the density of the life time T. Then, by a con-ditioning argument (similar to that in Section 15), the distribution of N_T is given by:

$$pr(N_T = j) \quad = \quad \int_0^\infty pr(N_T = j \mid T = t) \, f(t) \, dt$$

$$= \quad \int_0^\infty \frac{(\lambda t)^j}{j!} \, e^{-\lambda t} \, f(t) \, dt.$$

This is the required formula.

It is easy to see that the average number of arrivals is now:

$$E(N_T) \quad = \quad \lambda E(T)$$

in agreement with intuition; indeed, t in λt is now

replaced by $E(T)$ -- the average (see also a similar relation
in Section 15).

As an example, suppose that f is exponential with
parameter μ. Then

$$pr(N_T = j) \quad = \quad \int_0^\infty \frac{(\lambda t)^j}{j!} \ e^{-\lambda t} \mu e^{-\mu t} \ dt \quad = \quad \frac{\mu \lambda^j}{(\mu + \lambda)^{j+1}} \ ,$$

$$j = 0, 1, \ldots,$$

which is of course the geometric distribution with mean λ / μ
(see also Section 22).

Section 20: It Is Easy to Learn

We shall now consider a simple learning model in which
an individual learns to perform one action correctly. We
shall assume that when the action is performed, the individual's response may be either correct or incorrect. As time
progresses, one obtains a sequence of responses; if from
some time on, the correct responses prevail, we may consider
this as an indication that the individual learned to perform
the assigned task. If incorrect responses prevail, the individual did not learn anything. We assume that time flows
continuously, but tasks (i.e., actions) are performed successively at some aribitrary instants.

It will be convenient to denote responses by 0 and 1,
and classify them as

0 = correct response, 1 = wrong response.

Denote by Y_t the r.v. which indicates the response which

prevails at time t. Thus, $(Y_t = 0)$ is the event that at time t the correct response prevails; similarly $(Y_t = 1)$ is the event of incorrect response at time t. For simplification we shall say the system is in state 0 or in state 1 at t.

We shall be interested in the pr. of state j at time t, given that initially state i prevailed:

$$p_{ij}(t) = pr(Y_t = j \mid Y_0 = i),$$

where i and j can be only 0 and 1; $t \geq 0$. For example, $p_{10}(t)$ is the pr. of a correct response at time t, when initially the response was wrong. To make the problem more precise further assumptions are needed. We shall again consider what may happen within a short interval (t,t+h). We shall assume that during this interval, approximately:

pr. of correct response, when the wrong one prevailed at

t, is $p_{10}(h) = ah + \ldots$

pr. of wrong response, when the correct one prevailed at

t, is $p_{01}(h) = bh + \ldots$.

Coefficient a represents the learning rate; it would be desirable to have a large. Coefficient b represents the hindrance rate; it would be desirable to have b small. One could say that a is a measure of brightness, whereas b is a measure of dumbness.

To obtain equations for transition pr.'s $p_{ij}(t)$, we shall consider two contiguous intervals $(0,t)$ and $(t,t+h)$, and enumerate all possibilities of transitions. The following equations are self-evident -- they represent the "balance of pr.":

$$p_{00}(t+h) = p_{00}(t)p_{00}(h) + p_{01}(t)p_{10}(h)$$

$$= p_{00}(t)[1-bh] + p_{01}(t)ah + \ldots$$

$$p_{01}(t+h) = p_{00}(t)p_{01}(h) + p_{01}(t)p_{11}(h)$$

$$= p_{00}(t)bh + p_{01}(t)[1-ah] + \ldots$$

$$p_{10}(t+h) = p_{10}(t)p_{00}(h) + p_{11}(t)p_{10}(h)$$

$$= p_{10}(t)[1-bh] + p_{11}(t)ah + \ldots$$

$$p_{11}(t+h) = p_{11}(t)p_{11}(h) + p_{10}(t)p_{01}(h)$$

$$= p_{11}(t)[1-ah] + p_{10}(t)bh + \ldots$$

Hence, as in the previous section, putting $p_{ij}(t+h) - p_{ij}(t)$ on the left side, dividing by h, and passing to the limit $h \to 0$, one obtains the following system of differential equations:

$$\frac{d}{dt} p_{00}(t) = -bp_{00}(t) + ap_{01}(t) \qquad (1)$$

$$\frac{d}{dt} p_{01}(t) = bp_{00}(t) - ap_{01}(t) \qquad (2)$$

$$\frac{d}{dt} p_{10}(t) \;=\; -\,bp_{10}(t) \,-\, ap_{11}(t) \tag{3}$$

$$\frac{d}{dt} p_{11}(t) \;=\; bp_{10}(t) \,-\, ap_{11}(t) \tag{4}$$

with initial conditions:

$$p_{00}(0) = p_{11}(0) = 1, \qquad\qquad p_{01}(0) = p_{10}(0) = 0. \tag{5}$$

Observe also that necessarily,

$$p_{00}(t) + p_{01}(t) = 1, \qquad\qquad p_{10}(t) + p_{11}(t) = 1 \tag{6}$$

because possibilities of getting 0 or 1 are mutually exclusive. Note that the addition of equations (1)-(4) sidewise gives 0 = 0, as it should.

To solve the system, consider equation (1):

$$\frac{d}{dt} p_{00}(t) + bp_{00}(t) \;=\; ap_{01}(t) \tag{1}$$

and differentiate

$$\frac{d^2}{dt^2} p_{00}(t) + b\,\frac{d}{dt} p_{00}(t) \;=\; a\,\frac{d}{dt} p_{00}(t) \;=\; -a\,\frac{d}{dt} p_{00}(t)$$

using (6). So

$$\frac{d^2}{dt^2} p_{00}(t) + (a{+}b)\,\frac{d}{dt} p_{00}(t) \;=\; 0 \tag{1a}$$

Put $y(t) = \dfrac{d}{dt} p_{00}(t)$, so (1a) becomes

$$\frac{d}{dt} y(t) + (a+b) y(t) = 0 \tag{1b}$$

and its solution is

$$y(t) = Ae^{-(a+b)t}. \tag{1c}$$

Integrating (1c) again, one has

$$p_{00}(t) = B - \frac{A}{a+b} e^{-(a+b)t} \tag{1d}$$

where A and B are constants. These constants are determined from initial conditions (5):

$$p_{00}(0) = 1, \quad \frac{d}{dt} p_{00}(t)\Big|_{t=0} = ap_{00}(0) - bp_{00}(0) = -b.$$

Hence, letting t = 0 in (1c) and (1d):

$$A = -b, \quad 1 = B + \frac{b}{a+b} \quad \text{or} \quad B = \frac{a}{a+b}$$

one has:

$$\boxed{p_{00}(t) = \frac{a}{a+b} + \frac{b}{a+b} e^{-(a+b)t}} \tag{7}$$

Hence, from (6):

$$\boxed{p_{01}(t) = \frac{b}{a+b} - \frac{b}{a+b} e^{-(a+b)t}}. \tag{8}$$

Similar analysis applied to equation (3) yields:

$$P_{10}(t) = \frac{a}{a+b} - \frac{a}{a+b} e^{-(a+b)t} \tag{9}$$

$$P_{11}(t) = \frac{b}{a+b} + \frac{a}{a+b} e^{-(a+b)t} \tag{10}$$

Suppose that we do not know what the initial values at $t = 0$ are. Then, the initial distribution must be given:

$$\mathrm{pr}(Y_0 = 0) = \pi_0, \quad \mathrm{pr}(Y_0 = 1) = \pi_1, \quad \text{with} \quad \pi_0 + \pi_1 = 1.$$

The joint distribution is then

$$\mathrm{pr}(Y_0 = i, \ Y_t = j) = \pi_i P_{ij}(t), \quad \text{with} \quad i,j = 0,1.$$

Hence the distribution at t is:

$$\mathrm{pr}(Y_t = j) = \pi_0 P_{0j}(t) + \pi_1 P_{1j}(t), \quad \text{for} \quad j = 0,1.$$

Denote this pr. by $\pi_j(t)$. Then, in our case

$$\pi_0(t) = \frac{a}{a+b} + \frac{\pi_0 b - a \pi_1}{a+b} e^{-(a+b)t} \tag{11}$$

$$\pi_1(t) = \frac{b}{a+b} - \frac{\pi_0 b - a \pi_1}{a+b} e^{-(a+b)t} \tag{12}$$

Note that $\pi_0(t) + \pi_1(t) = 1$, for all t, as it should. Also, (11) and (12) reduce to π_0 and π_1, respectively, for $t = 0$.

More important, however, is the fact that the limiting distributions coincide:

$$\lim_{t \to \infty} p_{i0}(t) = \lim_{t \to \infty} \pi_0(t) = \frac{a}{a+b},$$

$$\text{(13)}$$

$$\lim_{t \to \infty} p_{i1}(t) = \lim_{t \to \infty} \pi_1(t) = \frac{b}{a+b}.$$

This means that as time progresses the system tends to an equilibrium -- that is states 0 and 1 are achieved with constant pr.'s, which are independent of the initial state. In other words, influence of the initial conditions wears out, and the system operates fluctuating between 0 and 1. For the learning model, this means that in the long run the pr. of correct responses is $\frac{a}{a+b}$; that of wrong responses is $\frac{b}{a+b}$.

Note that if a and b are chosen such that $\pi_0 b = \pi_1 a$, then $\pi_0(t)$ and $\pi_1(t)$ are constant for each t. Thus, for this special choice of initial distribution the system would behave as if it were already in equilibrium.

Indeed, the required condition can be satisfied iff the initial distribution (π_0, π_1) coincides with the limiting distribution (13).

It is of interest to plot graphs of $\pi_j(t)$. First of all note that:

$$\pi_0 < \frac{a}{a+b} \qquad \text{iff} \qquad \pi_0 b - a\pi_1 < 0;$$

$$\pi_0 > \frac{a}{a+b} \qquad \text{iff} \qquad \pi_0 b - a\pi_1 > 0.$$

Hence the graph of the pr. of correct response at time t is:

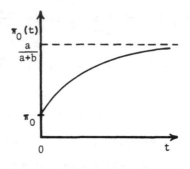

Fig 20.1

Probability of the correct response for $\pi_0 < a/(a+b)$.

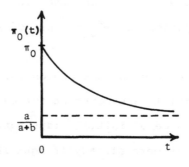

Fig 20.2

Probability of the correct response for $\pi_0 > a/(a+b)$.

Section 21:   On-Off Transitions

21.1.

Consider a single device (a bulb, machine, signal, neuron, etc.) which can only be in two states, say, on and off (busy-idle, good-bad, in-out, etc.).   For convenience, we shall denote

state "off" as state 0, and state "on" as state 1.

The device may spend some time in each state, and as time progresses busy and idle periods alternate. Let

X_0 be a life time representing the idle period (i.e., when state 0 prevails),

X_1 be a life time representing the busy period (i.e., when state 1 prevails).

The life time $X = X_0 + X_1$ represents a cycle of repetition, that is the time interval between two consecutive instants of the beginning of state 1 (interarrival time) or between two consecutive instant of the beginning of state 0 (termination time). See the Figure below.

We shall now assume that X_0 and X_1 are independent, and denote their density, d.f. and mean by f_0, F_0, μ_0 and f_1, F_1, μ_1, respectively. Similarly, let f, F and μ correspond to the cycle life time X.

As shown in Section 5, the density f is given by convolution of densities f_0 and f_1:

$$f(t) = \int_0^t f_0(s) f_1(t-s) \, ds$$

and clearly $\mu = \mu_0 + \mu_1$.

One could regard a sequence of cycles X as a renewal process (so called alternating process). Let u be the corresponding density of renewals. We have now the renewal equation

$$u(t) = f(t) + \int_0^t u(t-s) f(s) \, ds$$

and by the renewal theorem (see Section 17) the limiting rate is $1/\mu$. Thus the average number of arrivals, as well as of terminations, in time interval $(0,t)$ is approximately t/μ.

From now on, we shall assume that life times X_0 and X_1 are exponential with densities:

$$f_0(t) = be^{-bt}, \quad f_1(t) = ae^{-at}, \quad \text{for } t \geq 0, \quad a > 0, \quad b > 0.$$

Then:

$$f(t) = \int_0^t be^{-bs} ae^{-a(t-s)} \, ds = \frac{ab}{b-a}(e^{-at} - e^{-bt}) \geq 0$$

$$(t \geq 0)$$

for $a \neq b$; for $a = b$ see Example 1 in Section 5. The d.f. is

$$F(t) = 1 - \frac{1}{b-a}(be^{-at} - ae^{-bt}), \quad t \geq 0.$$

Clearly, $\mu = \frac{1}{a} + \frac{1}{b}$.

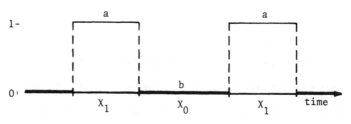

Fig 21.1

On-off transitions.

Now we shall consider another problem. Suppose that we wish to find the pr. that at some instant   t,   a device will be in state   0   or in state   1.   Let Y_t be a r.v. representing the state at   t.   Write again

$$P_{ij}(t) \quad = \quad pr(Y_t = j \mid Y_0 = i)$$

for the conditional pr. of state   j   at   t,   given that initially at   t = 0   state   i   prevailed. Here   i   and   j   can assume only values   0   and   1.

A moments reflection shows that these transition pr.'s $P_{ij}(t)$ are the same as those found in the previous Section 20.   However, we shall verify this statement by a different argument which further illustrates the connection between the life times we discussed earlier, and the r.v.'s representing a number of elements (here   0   and   1,   only).

Suppose that the device is initially in state   0   (that is "off").   We wish to find the pr. of it being in state 1 (i.e. "on") at   t.   Thus, the device is at   0   initially, and state   0   prevails until some instant   s   where the first transition to state 1 takes place. This occurs with density f_0 (termination of the idle state).   Then during the remaining time interval   t - s   transitions from state   1   to state 1 may occur, with pr. $P_{11}(t-s)$.   Hence, the total pr. is:

$$P_{01}(t) \quad = \quad \int_0^t f_0(s) P_{11}(t-s) \ ds.$$

If it is required that at   t   the device be in state 0,
when initially it has been in   0,   two possibilities must be
considered.   The device may stay in   0   up to time   t,   with
no change; this means that life time X_0 continues   beyond
t,   and the pr. of such an event is clearly $F_0^c(t) = e^{-bt}$.
The second possibility is that a device originally at   0,
changes to   1   and at some instant   s   returns for the first
time to state   0;   this corresponds to the cycle   X   with
density   f.   Then, during the remaining time interval   t‑s
some transitions from   0   to   0   may occur, with   pr.
$P_{00}(t-s)$.   The total pr. is therefore:

$$P_{00}(t) = F_0^c(t) + \int_0^t f(s) P_{00}(t-s) \ ds.$$

It can be easily verified by substitution that expressions
for the transition pr. $P_{ij}(t)$ found in the previous sec‑
tion satisfy the above equations.

By a similar argument, one obtains for the initial
state 1:

$$P_{10}(t) = \int_0^t f_1(s) P_{00}(t-s) \ ds$$

$$P_{11}(t) = F_1^c(t) + \int_0^t f(s) P_{11}(t-s) \ ds.$$

****

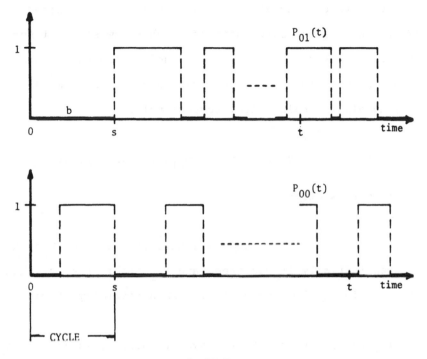

Fig 21.2

Transition probabilities.

21.2. Group of lines

Consider now a group of n lines, each of which may be busy or free, the change of state being governed by equations developed in this section (and in Section 20). Assume also that the individual lines act independently. We shall be now interested in the distribution of the number of busy lines in the group. Let Y_t be the number of busy lines at time t, with Y_0 being the initial number. We wish to find

$$P_{ij}(t) \;\; = \;\; pr(Y_t = j \mid Y_0 = i)$$

where now i and j range from 0 to n.

We previously treated the case $n = 1$. It is hoped that the use of the same notation in this new set will not confuse the issue. To avoid any misunderstanding, we shall use new symbols for the two transition pr.'s which we shall need, namely we shall write, respectively,

$$c_0(t) \quad \text{and} \quad c_1(t) \quad \text{for} \quad p_{01}(t) \quad \text{and} \quad p_{11}(t)$$

from subsection 21.1.

Thus, $c_0(t)$ is the pr. that a single line will be busy at t, when initially it was free, and $c_1(t)$ is the pr. that a single line will be busy at t, when initially it was busy, too.

Initially, at $t = 0$ we have i busy lines and $n - i$ free lines, and at time t we wish to have j busy lines. This change can be accomplished when originally free lines become busy, and originally busy lines remain busy, at least some of them. We must consider all possibilities.

Suppose, then that out of the $n - i$ free lines initially, k will become busy at t. We have here the Bernoulli trials with pr. of "success" being $c_0(t)$. Hence, the required pr. is:

$$q_i(k,t) \;=\; \binom{n-i}{k} [c_0(t)]^k [1-c_0(t)]^{n-i-k}$$

(see Section 11 and the somewhat analogous situation with the rabbit in Section 18.1).

Furthermore, as there should be j busy at t so
j-k must remain busy from the initial i busy. Thus,
again we have Bernoulli trials with pr. of "success" $c_1(t)$.
The required pr. is

$$r_i(j-k,t) = \binom{i}{j-k}[c_1(t)]^{j-k}[1-c_1(t)]^{i-j+k}.$$

Furthermore, transitions from free to busy and from
busy to busy on these lines, are independent, so in order
to get transition from i to j, we should multiply the
above pr.'s, and then add them through all possible intermediary k. Thus, our final result is:

$$p_{ij}(t) = \sum_k r_i(j-k,t)q_i(k,t), \qquad 0 \le t,$$

where actually the summation is taken over the range
$\max(j-i,0) \le k \le \min(n-i,j)$.

It is easy to see that the (conditional) average number of busy lines is:

$$E(Y_t \mid Y_0 = i) = (n-i)c_0(t) + ic_1(t) = \frac{nb}{a+b} + (1 - \frac{nb}{a+b})e^{-(a+b)t}$$

and tends to $\frac{nb}{a+b}$ when $t \to \infty$. (Evidently, when n = 1
the above results agree with those in Section 20 --con-

versely, the present situation of $n > 1$ corresponds to n independent learners.

Remark: Extension of analysis from the previous subsection makes the validity of the following equations connecting transition pr.'s with first passage distributions plausible:

$$p_{ij}(t) = \int_0^t f_{ij}(s) p_{jj}(t-s)\, ds \quad \text{for } i \neq j$$

$$p_{jj}(t) = e^{-q_j t} + \int_0^t f_j(s) p_{jj}(t-s)\, ds$$

where $q_j = (n-j)b + ja$, and f_{ij} is the density of time to first entrance to j (from i) and f_j is the density of the time of the first return to j.

****

Finally, it can be verified that the limiting distribution

$$\lim_{t \to \infty} p_{ij}(t) = P(j)$$

is independent of the initial state i, and is given by the binomial distribution with parameters n and $b/(a+b)$.

Section 22:   Queueing

Consider a "server" who performs a certain service for customers, like a clerk in a Post Office selling stamps, a

doctor seeing patients, a mechanic servicing cars; a car
washing facility, a bus stop with passengers waiting for a
bus, and so on. Customers may arrive in any fashion what-
soever. When the server is free, the incoming customer is
given service immediately. Customers arriving when the
server is busy, are put in a line and wait. When a server
completes the service, any customer from the waiting line
may be given access to the server, depending on the rules of
"queue discipline." The most typical is "first come first
serve," although other possibilities have been considered
too. For example, some priorities may be assigned to waiting customers, the "last come first served" being not
uncommon. On the other hand, customers may just scramble
for a server ("random servicing").

Of primary interest is the distribution of the customers present in the system, and the distribution of the waiting
time to obtain service. We shall consider this problem here
under very specific assumptions, which reasonably well represent the real life situation. The queueing system to be
discussed here is known as a simple queue.

Structure of the system:

(i)   only one server;

(ii) when the server is busy, a waiting line is formed
      which may be infinite in length;

(iii) customers are served on the "first come first
      served" basis (strict order servicing);

(iv)   no waiting line is possible when the server is
       free;

(v)   customers in the queue wait till serviced, then
       depart (no early departures prior to service).

Probabilistic properties of the simple queue are described
by the following assumptions:

(vi)   incoming customers form a Poisson input, with
       rate λ;

(vii) service times of customers are independent identi-
       cally distributed exponential life times, with
       common mean $1/\mu$;

(viii) service times and interarrival times are indepen-
       dent of each other.

   Let Y_t be a r.v. representing a number of customers
present in the system at the instant t, including a
customer being served, if any. Clearly, Y_t assumes values
0 (meaning no customers), 1 (necessarily a customer being
served), 2 (one served, one waiting), and so on. Thus,
$(Y_t = j)$ is the event that at time t there are exactly j
customers in the system; $j-1$ of them are waiting, if any.
The event $(Y_t = j)$ is expressed by saying that the system
is in state j at t.

   Customers arrive to the system, join the queue and
wait; then are served for the duration of their service time,
and depart afterwards. During their waiting and service,

other customers arrive. Thus, their total number Y_t
fluctuates as time t progresses.

We shall be interested in the conditional pr. that at
time t there are j customers in the system, when at time
0 there have been i customers (the initial state i at
$t = 0$ may correspond to the instant when observation began).
In symbols:

$$p_{ij}(t) = \text{pr}(Y_t = j \mid Y_0 = i), \qquad t \geq 0, \quad i,j = 0,1,2,\dots .$$

Thus, $p_{ij}(t)$ expresses the transition pr. from state i
at $t = 0$ to state j at time t. Note that the values
i and j are arbitrary. For example $p_{i0}(t)$ is the pr.
that although initially there were i customers, at time t
there is none (i.e., the server is free at t). The life
time representing distance between the instant when a free
server becomes engaged, and the instant when the server is
free again, is known as a busy period. Clearly busy periods
alternate with slack periods.

Assumptions stated above are used to derive equations
for the transition pr.'s $p_{ij}(t)$.

Fig 22.1
Waiting line with a single server.

For this purpose consider two contiguous intervals
(0,t)   and   (t,t+h)   where   h > 0   is small.   Suppose that
at   t = 0   there are already   i   customers present.   In
order to have   j   customers at   t + h,   one must consider
four possibilities:

(a)   There are   j - 1   customers at   t,   and one arrival
      occurred during   (t,t+h),

(b)   There are   j + 1   customers at   t,   and one termination
      occurred during   (t,t+h)

(c)   There are   j   customers at   t,   and no change occurred
      during   (t,t+h),

(d)   more than one change occurred during   (t,t+h).

The arrival of a new customer means that the interarrival period terminates during   (t,t+h)   and we have seen
in Section 3 that for the exponential life time this occurs
with pr. λh,   approximately; indeed we have here the Poisson
input discussed in Section 19.   Clearly, the arrival of a
new customer causes an increase in the value of Y_t by   1,
no matter how many customers are already in the system at
the instant of arrival.   Thus, the pr. of contingency (a)
is approximately:

$$P_{j-1,j}(h) = \lambda h.$$

Similarly, the termination during (t,t+h) of a service time in progress occurs with pr. μh, approximately, because service times are exponential. Clearly the termination of service causes a decrease in the value of Y_t by 1, and this is the only way customers can depart. Thus the conditional pr. of contigency (b) is approximately:

$$P_{j+1,j}(h) = \mu h.$$

By assumptions (vi) - (viii), the pr. of more than one transition during (t,t+h) is negligible. Hence, the pr. of contingency (c) is approximately

$$P_{jj}(h) = 1 - \lambda h - \mu h.$$

In terms of pr.'s, the above contingencies can be expressed by the "balance equation":

$$P_{ij}(t+h) = P_{i,j-1}(t)P_{j-1,j}(h) + P_{i,j+1}(t)P_{j+1,j}(h)$$

$$+ P_{ij}(t)P_{jj}(h) + \ldots .$$

Substituting the expressions found above, the balance equation becomes (up to approximation):

$$P_{ij}(t+h) = P_{i,j-1}(t)\lambda h + P_{ij}(t)(1 - \lambda h - \mu h)$$

$$+ P_{i,j+1}(t)\mu h + \ldots \qquad \text{for small } h.$$

Hence

$$\frac{P_{ij}(t+h) - P_{ij}(t)}{h} = P_{i,j-1}(t)\lambda - P_{ij}(t)(\lambda+\mu)$$

$$+ P_{i,j+1}(t)\mu + \ldots .$$

Passing to the limit as $h \to 0$, one obtains a difference-differential equation:

$$\boxed{\frac{dp_{ij}(t)}{dt} = \lambda p_{i,j-1}(t) - (\lambda+\mu)p_{ij}(t) + \mu p_{i,j+1}(t)} \quad (*)$$

which is valid for all i and for $j = 1,2,\ldots$, and for all $t \geq 0$. For $j = 0$, terminations are impossible, so by the similar reasoning

$$\boxed{P_{i0}(t+h) = P_{i0}(t)(1 - \lambda h) + P_{i1}(t)\mu h + \ldots}$$

which leads to:

$$\boxed{\frac{dp_{i0}}{dt}(t) = -\lambda p_{i0}(t) + \mu p_{i1}(t)}. \quad (**)$$

Note that i and j assume all values $0,1,2,\ldots$ and it is immaterial whether i is larger or smaller than j. Observe also that assumption (iii) did not intervene. It may be of interest to note that in the special case when $\mu = 0$, the above equations reduce to those of the Poisson process discussed in Section 19. It is the presence of terminations which results in the fluctuation of values of Y_t, and not merely increasing as in the Poisson case.

The initial conditions are expressed by:

$$P_{ii}(0) = 1, \quad P_{ij}(0) = 0 \quad \text{for} \quad i \neq j. \qquad (***)$$

Although the above infinite system of equations can be solved explicitly, the solution is rather complicated and therefore not very informative. For practical applications is of greater interest to see what is the solution in equilibrium, that is after a sufficiently long time when "steady state" conditions prevail. The important fact is that such a solution exists, and is independent of initial conditions (that is the effect of initial conditions wears out as time progresses), provided a rather mild and obvious restriction is imposed. Moreover, this equilibrium solution has very simple and familiar form.

Thus, we shall assume that the following limit exists for every $j = 0,1,\ldots$

$$\lim_{t \to \infty} P_{ij}(t) = P(j)$$

and is independent of the initial state i, and that P is proper distribution:

$$\sum_{j=0}^{\infty} P(j) = 1.$$

It is intuitively clear that as $p_{ij}(t)$ tends to a constant, its derivative would tend to 0:

$$\lim_{t \to \infty} \frac{dp_{ij}}{dt}(t) = 0.$$

Passing to the limit with $t \to \infty$ in equations (*) and (**), one obtains the so called "equilibrium equations":

$$0 = -\lambda P(0) + \mu P(1), \qquad \text{for} \quad j = 0$$

$$0 = \lambda P(j-1) - (\lambda + \mu)P(j) + \mu P(j+1), \qquad \text{for} \quad j = 1,2,\ldots$$

One can solve this system very easily. Consider first equations for $j = 0$ and $j = 1$:

$$0 = \lambda P(0) - (\lambda + \mu)P(1) + \mu P(2) \qquad (\text{for} \quad j = 1).$$

Substituting from equation for $j = 0$, one has from the above that

$$0 = -\lambda P(1) + \mu P(2).$$

Proceeding in the same manner for $j = 2,3,\ldots$, one reduces the above system to a much simpler form:

$$0 = -\lambda P(j) + \mu P(j+1) \qquad \text{for} \quad j = 0,1,\ldots$$

or

$$P(j+1) = (\lambda/\mu)P(j) = \rho P(j) \qquad \text{for} \quad j = 0,1,\ldots$$

where $\rho = \lambda/\mu$ is known as the queueing intensity. Note that ρ is the ratio of the average service time and the average interarrival time (see Example 12.4).

Hence:

$$P(1) = \rho P(0)$$
$$P(2) = \rho P(1) = \rho^2 P(0)$$

so by induction

$$P(j) = \rho^j P(0) \qquad \text{for } j = 0,1,\ldots .$$

This is almost what we want, except that $P(0)$ is unknown. To find $P(0)$ we must use the fact that all $P(j)$ add to 1, so

$$1 = \sum_{j=0}^{\infty} P(j) = P(0) \sum_{j=0}^{\infty} \rho^i = P(0)(1-\rho)^{-1}$$

provided $\rho < 1$, because otherwise the geometric series would not converge. This is the required condition for existence of the steady state solution $P(j)$. It now assumes the final form

$$\boxed{P(j) = (1-\rho)\rho^j \qquad \text{for } j = 0,1,\ldots}$$
.

This is the geometric distribution (see Section 13). Clearly $E(Y_t) = \rho/(1-\rho)$.

Section 23:   Waiting Time

We shall now continue the discussion of the queueing system described in the preceding section. We have already found the distribution of the number of customers in the system at an arbitrary instant t in equilibrium, that is,

$$P(j) \quad = \quad pr(Y_t = j).$$

This is the geometric distribution with parameter ρ.

We shall now consider the waiting time W for an arbitrary customer, and we shall determine its distribution:

$$W(t) \quad = \quad pr(W \le t), \qquad t \ge 0.$$

That is, W(t) is the d.f. of the life time W.

In order to achieve this, it is most convenient to follow an individual customer arriving at some instant. If the server is free, this customer is given service immediately, and his waiting time is zero, with positive pr. We have here a new phenomenon which we did not see earlier, namely the possibility of a life time W assuming a single value (zero) with pr. not equal to zero.

If the server is busy, the customer joins the queue and waits until his turn comes. Assumption (iii) on strict order of service is essential here. Consider now

$G_n^c(t) \quad = \quad$ the pr. that a customer who on his arrival

has met n - 1 waiting customers and became

the n-th in the queue, has not yet received
service up to time t (for n = 1,2,...
and t ≥ 0).

This is in fact the complementary d.f. of the waiting time,
conditional that at the instant of arrival there will be
exactly n customers in the system, excluding that new
arrival. Hence, necessarily n = 1,2,... .

It is intuitively clear, and it may be verified
analytically, that the pr. of having exactly n customers
at instants of arrival should coincide with the pr. of
exactly n customers at an arbitrary instant in equilibrium.
That is P(n), found above. Hence, taking into account the
values of n (see Section 3 on conditioning), one obtains
the formula for the complementary d.f. of the waiting time:

$$W^C(t) \quad = \quad \sum_{n=1}^{\infty} P(n)G_n^C(t) \quad = \quad (1-\rho) \sum_{n=1}^{\infty} \rho^n G_n^C(t).$$

This is the pr. that the waiting time is larger than t.

It is now required to find functions G_n^C. Consider
again two intervals (0,t) and (t,t+h) where h > 0 is
small. Observe that by assumption of the strict order of
service, customers who arrive after our customer, do not
influence his waiting. Thus, during (t,t+h) only two kinds
of transitions can occur:

(a) termination of the actual service, with pr. μh (approximately), so there will be n - 1 customers in the system

(b)   no change in the number of customers (who may influence
       the waiting time); this occurs with pr. $1 - \mu h$, assuming
       that more than one transition has negligible pr.

Thus for $n > 1$, one has the balance equation

$$G_n^C(t+h) \;=\; \mu h G_{n-1}^C(t) + (1-\mu h) G_n^C(t)$$

whereas for $n = 1$ the balance equation is simply:

$$G_1^C(t+h) \;=\; (1 - \mu h) G_1^C(t).$$

Hence, as before, one obtains the difference-differential
equations for G_n^C:

$$
\begin{aligned}
\frac{dG_1^C}{dt}(t) &\;=\; -\mu G_1^C(t), \\[2ex]
\frac{dG_n^C}{dt}(t) &\;=\; \mu G_{n-1}^C(t) - \mu G_n^C(t), \qquad \text{for } n = 2,3,\ldots .
\end{aligned}
$$

and the initial condition is clearly $G_n^C(0) = 1$.

   We must now solve this system. It is a rather curious
fact that we do not need this explicit form of G_n^C. Indeed,
looking at the formula for $W^C(t)$, we only need a function
of the form:

$$\varphi(z,t) \;=\; \sum_{n=1}^{\infty} z^n G_n^C(t), \qquad \text{for } 0 \le z < 1$$

because then:

$$W^c(t) \ = \ (1-\rho)\varphi(\rho,t).$$

Now, multiply each equation for G_n^c by z^n, and then add all equations for all n. One then has

$$\frac{d\varphi}{dt} \ = \ \mu z\varphi - \mu\varphi \ = \ -(1-z)\mu\varphi$$

which is a single differential equation. The initial condition is $\varphi(z,0) = \frac{z}{1-z}$. To solve it write

$$\frac{d\varphi}{\varphi} \ = \ -(1-z)\mu \ dt.$$

Integrating one has

$$\varphi \ = \ Ce^{-(1-z)\mu t}.$$

From initial conditions, the constant is found to be $C = \frac{z}{1-z}$. Hence, the complete solution is:

$$\varphi(z,t) \ = \ z(1-z)^{-1}e^{-(1-z)\mu t}.$$

(We have seen an analogous procedure in Section 20.) Hence, substituting $z = \rho$, we get our final solution:

$$W^c(t) \ = \ \rho e^{-(1-\rho)\mu t}, \qquad \text{for } t \geq 0.$$

This is the required pr. that the waiting time will be longer than t. Remarkably simple formula for so much work!
Observe that for t = 0, it gives ρ,   the pr. that the customer need not wait (i.e., that his waiting time is zero) is

$$W(0) = 1 - \rho.$$

This is a remarkable result. Recall that we must have $\rho < 1$.

****

Remark: It is of interest to mention the alternative way of solution which also leans on other problems we discussed earlier.

Consider again the arriving customer who joined the line and became the n-th in the queue. There are n-1 customers in front of him in the line, and one customer being served. Total n life times -- service times, which must terminate before our customer is admitted to the server. Thus his waiting time may be represented as the sum:

$$W_n = X_1 + \ldots + X_{n-1} + X_n$$

By our assumptions, the first n-1 life times are exponential with parameter μ; the last one is a portion of a service time since the instant of arrival of our customer until the termination of a service (of a customer being served when our customer joined the line). By considerations from Section 4, this remaining life time is also an exponential with the same μ. Thus, all life times forming the waiting time W_n are i.i.d.

Thus, we have here the situation discussed in Section 16 (with W_n corresponding to S_n), so the density of W_n is

$$g_n(t) = \frac{(\mu t)^{n-1}}{(n-1)!} e^{-\mu t} \mu, \qquad n = 1, 2, \ldots, \quad t \geq 0.$$

Hence, the explicit form of the complementary d.f. of W_n is:

$$G_n^C(t) = \sum_{v=0}^{n-1} \frac{(\mu t)^v}{v!} e^{-\mu t}.$$

This is the explicit form of the solution of the difference-differential equations for G_n^C. (Constrast this with the analogous equations for the Poisson distribution in Problem 2).

Substituting this expression for $G_n^C(t)$ into the formula for $W^C(t)$, one has:

$$W^C(t) = (1-\rho) \sum_{n=1}^{\infty} \rho^n \sum_{v=0}^{n-1} \frac{(\mu t)^v}{v!} e^{-\mu t}.$$

Interchanging the order of summation, one obtains the same expression for $W^C(t)$ found above.

****

Return now to the distribution of waiting time.   It has
the form:

$$W(t) \ = \ 1 \ - \ \rho e^{-(1-\rho)\mu t}, \qquad t \geq 0 .$$

Recall that this is the pr. that the waiting time is   t   or
less.

Differentiation would yield

$$w(t) \ = \ \rho(1-\rho)\mu e^{-(1-\rho)\mu t}$$

which however is not the density, because integrating back
one only gets:

$$\int_0^t w(s) \ ds \ = \ \int_0^t \rho(1-\rho)\mu e^{-(1-\rho)\mu s} \ ds \ = \ \rho[1 - e^{-(1-\rho)\mu t}] .$$

Strangely enough, this does NOT equal $W(t)$.   This effect is
due to the discontinuity of $W(t)$ at $t = 0$,   so there is
no derivative at $t = 0$.

Fortunately, the mean waiting time can be computed
easily from the formula:

$$E(W) \ = \ \int_0^\infty W^c(t) \ dt \ = \ \rho \int_0^\infty e^{-(1-\rho)\mu t} \ dt \ = \ \rho(1-\rho)^{-1}\mu^{-1} .$$

It is convenient to compare the ratio of the mean waiting time $E(W)$ and the mean service time $E(X) = 1/\mu$:

$$\frac{E(W)}{E(X)} = \frac{\rho}{1-\rho} \; .$$

This ratio increases with ρ, and equals 1 for $\rho = \frac{1}{2}$. Note that this ratio is equal to the average number of customers in the system, $E(Y_t)$.

Remark: For still another approach to the waiting time see Problem 28.

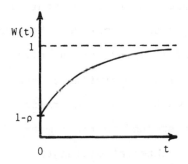

Fig 23.1

Waiting time distribution.

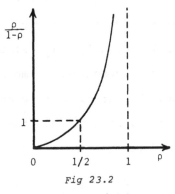

Fig 23.2

Ratio $E(W)/E(X)$.

Section 24: Birth Right

Consider a population of material objects or living organisms, and suppose that we are only interested in observing the increase of the size of this population (that is "births" in the population). For example, we may observe a growth of a colony of bacteria, or of a herd of sheep; we may investigate the spread of epidemics, arrivals of customers, accumulation of claims, orders and bills, etc. The situation is actually the same as that of input discussed in Section 19, except that now we shall impose more general requirements (of which the Poisson input will be a specific case). Basically, we shall consider the situation when the rate of births (input) depends on the size of this population. Our analysis will be similar to that in Section 22 (but we disregard "deaths," and generalize the input).

We shall be interested in the number of births occurring within the time interval of length t (and we shall assume as before that it is the length of the interval which matters, but not its placement on the time axis). Let Y_t be a r.v. representing a number of births until time t. Here Y_0 is the initial number of births (at the instant of starting observations). Clearly, Y_t assumes values $0,1,2,\ldots,$ and can only increase as time progresses. Hence $(Y_t = j)$ is the event that at time t exactly j births have been observed (the system is said then to be in state j).

We shall try to determine the transition pr.'s:

$$p_{ij}(t) \;=\; pr(Y_t = j \mid Y_0 = i), \quad 0 \le t$$

that at time t the system is in state j (j births
occurred), given that initially it was in state i (the
initial i births). We see immediately that $p_{ij}(t) = 0$
for $i > j$, so only the case of $j \geq i$ is of interest.

The family $(Y_t, 0 \leq t < \infty)$ is called a <u>birth</u> <u>process</u>,
provided the following assumptions hold:

A1: the conditional pr. that during an interval (t,t+h)
     where $h > 0$ is small, a birth occurs, given that
     at the beginning of the interval the system was in
     state i, is approximately equal to $\lambda_i h$.

A2: the conditional pr. of more than one birth during
     (t,t+h) is negligible.

Physically this implies that the number of births can
only increase in unit steps, with the pr. of a new birth independent of time but dependent on the size of the population,
and proportional to the interval length h. The factor of
proportionality λ_i is called the intensity or <u>birth</u> <u>rate</u>.

In order to derive equations for transition pr.'s,
consider two contiguous intervals (0,t) and (t,t+h) where
$h > 0$ is small. Suppose that at time t = 0 there are
already i births recorded. In order to have j births
at time t + h, one must consider three possibilities:

(a) there have been j - 1 births at t, and one birth
     occurred during (t,t+h),

(b)   there have been   j   births at   t,   and no birth

       occurred during   (t,t+h),

(c)   more than one birth occurred during   (t,t+h).

With the state $j-1$ prevailing at time   t,   the conditional

pr. of a new birth during   h   depends on $j-1$ and   h,   so

the conditional pr. of contingency a) is approximately:

$$P_{j-1,j}(h) \quad = \quad \lambda_{j-1}h.$$

On the other hand, the conditional pr. of contingency (b) is

approximately

$$P_{jj}(h) \quad = \quad 1 - \lambda_j h$$

because $\lambda_j h$ is the pr. of one birth.   By A2, the pr. of

contingency (c) is negligible.   In terms of pr.'s, the above

contingencies can be expressed by the "balance equation":

$$P_{ij}(t+h) \quad = \quad P_{i,j-1}(t)P_{j-1,j}(h) + P_{ij}(t)P_{jj}(h) + \ldots \, .$$

Substituting the expressions found above, the balance equa-

tion becomes (up to approximation) for small   h:

$$P_{ij}(t+h) \quad = \quad P_{i,j-1}(t)\lambda_{j-1}h + P_{ij}(t)(1-\lambda_j h) + \ldots \, .$$

Hence:

$$\frac{P_{ij}(t+h) - P_{ij}(t)}{h} \quad = \quad P_{i,j-1}(t)\,\lambda_{j-1} - P_{ij}(t)\lambda_j + \ldots \, .$$

Passing to the limit as $h \to 0$, one obtains a difference-differential equation:

$$\frac{dp_{ij}}{dt}(t) = p_{i,j-1}(t)\lambda_{j-1} - p_{ij}(t)\lambda_j \qquad (*)$$

which is valid for all i and j such that $j > i$ and $i = 0,1,2,\ldots$, and for all $t \geq 0$.

For $j = i$, only the second contingency can occur, so by similar reasoning:

$$p_{ii}(t+h) = p_{ii}(t)(1-\lambda_i h) + \ldots$$

which leads to:

$$\frac{dp_{ii}}{dt}(t) = -p_{ii}(t)\lambda_i . \qquad (**)$$

The initial conditions are expressed by:

$$p_{ii}(0) = 1, \quad p_{ij}(0) = 0 \qquad \text{for } j > i. \qquad (***)$$

Clearly, these equations are similar to equations stated in Section 22, and reduce to equations for the Poisson Process when all λ_i are equal. In contrast, with equations in Section 22, the birth equations derived here, can be easily solved.

We shall now find this solution and examine its properties. For this purpose we shall use Laplace transforms
defined in Section 7. Write:

$$u_{ij}(\alpha) = \int_0^\infty e^{-\alpha t} p_{ij}(t)\, dt, \qquad \alpha > 0.$$

We shall proceed by induction. First, multiply both sides
of equation (**) by $e^{-\alpha t}$ and integrate. Remembering the
initial condition 1, one finds (as noted in Section 7) that

$$\alpha u_{ii}(\alpha) - 1 = -u_{ii}(\alpha)\lambda_i$$

or

$$u_{ii}(\alpha) = \frac{1}{\alpha + \lambda_i}$$

which yields the explicit solution

$$p_{ii}(t) = e^{-\lambda_i t}.$$

Taking the transform of equation (*), and remembering that
the initial conditions vanish for $i \neq j$ one has similarly:

$$\alpha u_{ij}(\alpha) = u_{i,j-1}(\alpha)\lambda_{j-1} - u_{ij}(\alpha)\lambda_j$$

or

$$u_{ij}(\alpha) = u_{i,j-1}(\alpha)\frac{\lambda_{j-1}}{\alpha + \lambda_j}, \qquad \text{for } j > i.$$

As the product of Laplace transforms is the transform of convolution, the above relation becomes

$$p_{ij}(t) = \lambda_{j-1} \int_0^t e^{-\lambda_j(t-s)} p_{i,j-1}(s) \, ds, \qquad j > i.$$

This may be used to calculate $p_{ij}(t)$ recursively. It is more convenient, however, to work with Laplace transforms. The above recurrence relations for $u_{ij}(\alpha)$ can be solved directly (see a similar argument in Section 22) to give:

$$u_{ij}(\alpha) = \frac{1}{\alpha+\lambda_j} \frac{\lambda_{j-1}}{\alpha+\lambda_{j-1}} \cdots \frac{\lambda_i}{\alpha+\lambda_i} \qquad \text{for } j > i.$$

And this is our final solution. All we need do, is to substitute for the coefficients $\lambda_i, \ldots, \lambda_j$ and to invert the Laplace transform to get $p_{ij}(t)$. For example, if all $\lambda_i = \lambda$, then

$$u_{ij}(\alpha) = \frac{1}{\alpha+\lambda} \left(\frac{\lambda}{\alpha+\lambda}\right)^{j-i}, \qquad j \geq i$$

which is the Laplace transform of the Poisson distribution.

However, a lot of useful information can be obtained directly from the Laplace transform, without finding $p_{ij}(t)$ explicitly. For example, recall from Section 7 that

$$\lim_{t \to \infty} p_{ij}(t) = \lim_{\alpha \to 0} \alpha u_{ij}(\alpha)$$

and it follows from the above expression for $u_{ij}(\alpha)$ that for all i and j:

$$\lim_{t \to \infty} p_{ij}(t) = 0$$

(in contrast with the limiting distribution in Section 22).
This means that as time increases, the pr. of having any
finite number of births goes to zero (so we may expect that
the number of births will increase indefinitely).

Persuing this observation further, consider the expres-
sion $u_{ij}(\alpha)\lambda_j$ and note that it is the product of Laplace
transforms of exponential distributions. In particular, the
exponential form of $p_{ii}(t)$ means that the length of time
T_i the system spends in state i is exponentially distributed with the parameter λ_i. It then follows that the life
time T_j (i.e., time spent in state j) has exponential
distribution with the parameter λ_j, hence mean $1/\lambda_j$, for
each $j \geq i$:

$$pr(T_j > t) = e^{-\lambda_j t}.$$

Consider now the total time T spent in all states --
the life time of the birth process -- defined by

$$T = \sum_{j=i}^{\infty} T_j.$$

Taking expectations, the average total life time is clearly

$$E(T) = \sum_{j=1}^{\infty} \frac{1}{\lambda_j}.$$

Intuitively, we would expect $E(T)$ to be infinite, like in

the Poisson process. It is strange, however, that $E(T)$ may be finite in some cases (depending on the choice of λ_j).

Indeed, T itself is a rather strange life time (it is a duration of the whole birth process) and it is natural to expect that it will be infinite (and then obviously $E(T)$ would be infinite, too). Well, not quite! Let us write

$$F^C(t) = pr(T > t \mid Y_0 = i)$$

for the (complementary) d.f. of T, conditional on the initial state i (we shall drop i from the notation for simplicity). Taking the limit as $t \to \infty$, $F^C(\infty)$ is the pr. that T is infinite, and $F(\infty)$ is the pr. that T is finite.

It can be shown (by a renewal argument similar to that in Section 16) that

$$F^C(t) = \sum_{j=i}^{\infty} P_{ij}(t).$$

In view of the above remarks, as well as by the total probability argument, we would expect the sum to be equal to 1. This would correspond to T being infinite with pr. 1, i.e., $F(t) = 0$ for all finite t and $F(\infty) = 0$ or $F^C(\infty) = 1$.

But curious thing happen! There are cases (depending on the choice of λ_j) that

$$\sum_{j=i}^{\infty} P_{ij}(t) \leq 1$$

so $F(t)$ need not be zero for finite t, and consequently

F(t)   represents the distribution of   T,   with   F(∞)   being
the pr. that   T   is finite.

    But   T   is the total time for the process to reach
infinitely many births, so   T   finite means that infinitely
many births will occur in finite time. Some kind of explosion! It can be shown (by analytic methods which we cannot
discuss here) that   F(∞)   can be either   0   or   1   only, and
that:

$F^c(\infty) = 1$ (T   infinite with pr. 1) if and only if $E(T) = \infty$

$F^c(\infty) = 0$ (T   finite with pr. 1)   if and only if $E(T) < \infty$.

This is a very interesting mathematical result. We already
noted that for the Poisson input   T   is infinite with pr. 1
and the series for   E(T)   obviously diverges.

    Similarly, if we take $\lambda_j = j\lambda$ for $j \geq 1$,   and remember
that

$$\sum_{j=1}^{\infty} \frac{1}{j} = \infty,$$

then again $E(T) = \infty$ (for every   i),   so   T   is infinite
with pr. 1.

    On the other hand, taking $\lambda_j = j^2\lambda$ for $j \geq 1$ and
using the fact that

$$\sum_{j=1}^{\infty} \frac{1}{j^2} = \frac{\pi^2}{6} < \infty,$$

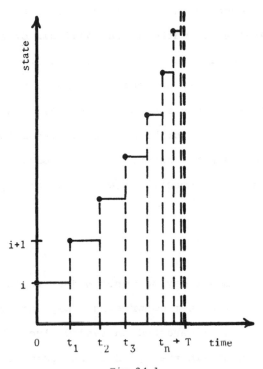

Fig 24.1

Birth process.

one finds that $E(T) < \infty$, so T is finite with pr. 1.
Thus, the explosion does occur, indeed.

On this thoughtful note, we close our discussion of
life times.

Section 25: Birth and Death

A dilligent reader of this chapter has noticed no doubt
a common pattern in our discussion of transition probabilities
$p_{ij}(t)$. Indeed, it is really cumbersome to repeat each time

a problem is presented the same line of reasoning. Fortunately the character of mathematical analysis offers an economy
of thought by singling out this common pattern and study its
properties. In this way we obtain a general model of importance which embraces many special cases. All that is needed
is to check whether the problem on hand fits into our general
model -- if so, then the solution of the model will automatically hold in our special case, and we are spared lengthy
derivations.

25.1.

Our general model may be described as follows. We have
a certain physical system comprising equipment, people or
animals, and we are interested in random fluctuations of a
number of components in the system. For example, we may look
at the number of operating machines in a work shop, the number of patients in the waiting room, the number of fish in
a pond, the number of students in a class, etc. To be more
definite, we shall say that we wish to investigate the number of customers in the system. But we shall leave to the
reader's imagination what is meant by a system and what is
meant by a customer (if you are still at a loss, think about
the queueing system in Section 22). Clearly, the number of
customers is $0, 1, \ldots$, etc. until say, we allow a finite
number (or the number may be infinite). We shall say that
the system is in the state i when there are exactly i
customers present. In our picturesque language, we shall
regard the increase in the number of customers as a birth,

and decrease as a death. Hence the name "birth and death"
process.

In probabilistic formulation, we shall consider a family
of r.v.'s (discrete life times) Y_t for $0 \leq t < \infty$, where
each Y_t represents the state of the system at the instant
t. Thus, $(Y_t = i)$ is the event that at time t, there are
exactly i customers in the system (i.e., the system is in
state i). We shall consider transition probabilities:

$$p_{ij}(t) = pr(Y_t = j \mid Y_0 = i) \qquad \text{for} \quad i,j = 0,1,\ldots, \quad t \geq 0$$

that at time t the system is in state j, given that it
was initially in state i. We shall impose an assumption of
homogeneity which means that this pr. $p_{ij}(t)$ is the same
for any pair Y_{t+s} and Y_s, irrespective of s.

We shall now impose a rather powerful assumption (which
we tacitly used in this chapter) that our process has the
Markov property, namely that the transition pr.'s satisfy the
"balance equation":

$$p_{ij}(t+h) = \sum_k p_{ik}(t) p_{kj}(h), \qquad t \geq 0, \quad h \geq 0 .$$

The meaning of this equation can be explained as follows.
Suppose that at time 0 the system is in state i, but at
time $t+h$ it should be in state j. How does it go from
i to j ? First it moves to state k at time t, and
then from k to j during the remaining time h. The associated pr.'s are $p_{ik}(t)$ and $p_{kj}(h)$, respectively, and

their product $p_{ik}(t)p_{kj}(h)$ is the pr. of the path from i through k to j. Summation over all k (all paths) yields the required pr. of transition. The Markov property is expressed by the fact that $p_{kj}(h)$ is independent of the state i at time 0. In other words, the transition from a state to another state depends on these states only, and not on the way the first state has been reached.

We also have the obvious requirements:

$$p_{ij}(t) \geq 0, \quad \sum_j p_{ij}(t) = 1, \quad \lim_{t \to 0} p_{ij}(t) = \begin{cases} 0 & \text{if } i \neq j \\ 1 & \text{if } i = j \end{cases}.$$

The family $(Y_t, \ 0 \leq t < \infty)$ is called a <u>birth</u> <u>and</u> <u>death</u> <u>process</u> if the following limits exist for each j:

$$\lim_{h \to 0} \frac{p_{j,j+1}(h)}{h} = \lambda_j, \qquad \lim_{h \to 0} \frac{p_{j,j-1}(h)}{h} = \mu_j,$$

$$\lim_{h \to 0} \frac{p_{jr}(h)}{h} = 0 \qquad \text{for } r \neq j-1, j, j+1$$

and

$$\lim_{h \to 0} \frac{1 - p_{jj}(h)}{h} = \lambda_j + \mu_j.$$

Probabilistically, these assumptions mean that:

A1: the conditional pr. that during an interval (t,t+h) where h > 0 is small, a transition (birth) to state

$j + 1$ occurs, given that at t the system was in state j, is

$$P_{j,j+1}(h) = \lambda_j h \quad \text{approximately;}$$

A2: the conditional pr. that during an interval $(t, t+h)$ where $h > 0$ is small, a transition (death) to state $j - 1$ occurs, given that at t the system was in state j, is

$$P_{j,j-1}(h) = \mu_j h \quad \text{approximately;}$$

A3: the conditional pr. of more than one transition during $(t, t+h)$ is negligible, and of no transition is

$$1 - P_{jj}(h) = (\lambda_j + \mu_j)h \quad \text{approximately.}$$

Physically, this implies that the system changes (during h) from a state to its next neighbor only, or no change takes place. Thus we have births and deaths by a unit only, the multiple transitions being excluded; see Fig. 25.1. The constants λ_j and μ_j are called the birth and death intensities or <u>rates</u>. Clearly $\mu_0 = 0$ and if the number of states is finite, say n, then $\lambda_n = 0$.

In order to derive equations for the transition pr.'s, we shall use our balance equation for $P_{ij}(t+h)$. Separating the three terms corresponding to $k = j - 1, j, j + 1$, we have:

$$p_{ij}(t+h) = p_{i,j-1}(t)p_{j-1,j}(h) + p_{ij}(t)p_{jj}(h)$$

$$+ p_{i,j+1}(t)p_{j+1,j}(h) + \sum p_{ik}(t)p_{kj}(h)$$

hence:

$$\frac{p_{ij}(t+h) - p_{ij}(t)}{h} = p_{i,j-1}(t)\frac{p_{j-1,j}(h)}{h} - p_{ij}(t)\frac{1 - p_{jj}(h)}{h}$$

$$+ p_{i,j+1}(t)\frac{p_{j+1,j}(h)}{h} + \sum p_{ik}(t)\frac{p_{kj}(h)}{h}$$

where the above sums are extended over $k \neq j-1$, j, $j+1$.

Passing to the limit with $h \to 0$, we obtain the basic birth and death equation:

$$\boxed{\frac{dp_{ij}}{dt}(t) = p_{i,j-1}(t)\lambda_{j-1} - p_{ij}(t)(\lambda_j + \mu_j) + p_{i,j+1}(t)\mu_{j+1}} \quad (*)$$

for $i \geq 0$ and $j > 0$. For $j = 0$, no death can occur, so the same reasoning yields:

$$\frac{dp_{i0}}{dt}(t) = -p_{i0}(t)\lambda_0 + p_{i1}(t)\mu_1,$$

and if the number of states is finite, then (no birth possible) the last equation is :

$$\frac{dp_{in}}{dt}(t) = p_{i,n-1}(t)\lambda_{n-1} - p_{in}(t)\mu_n.$$

The initial conditions are expressed by:

$$p_{ii}(0) = 1, \quad p_{ij}(0) = 0 \quad \text{for } i \neq j.$$

It is now clear that our equation (*) is the generalization
of equations encountered earlier. When all $\lambda_j = \lambda$ and
all $\mu_j = 0$, we get the Poisson process, and when $\mu_j = 0$
for all j we obtain the birth process (Section 24). In
Section 20 there are only two states $(n = 1)$ and $\lambda_0 = b$,
$\mu_0 = 0$, $\lambda_1 = 0$, $\mu_1 = a$. Finally, in Section 22 we have
$\lambda_j = \lambda$ and $\mu_j = \mu$. It is also clear that our considerations of all possible contingencies in those sections are
now embraced by our single derivation here from the balance
equation.

Now, for the practical applications, all we need to do
is to specify the rates λ_j and μ_j, and to substitute
them in the ready-made single equation (*). You see the
saving of labor achieved in this manner! The choice of coefficients λ_j and μ_j depends of course on the problem on
hand. Earlier sections in this chapter provide ample
examples, but we shall work out a few more soon. Incidently,
note that we considered transitions during h, following
the instant t -- for this reason equation (*) is conveniently
called the forward equation.

Remark. It may happen that for some state i one has identically in t that $p_{ii}(t) = 1$, so consequently $p_{ij}(t) = 0$ for
all $j \neq i$. This means that no transitions are possible
from i—such a state is said to be absorbing. Clearly, for
absorbing state $\lambda_i + \mu_i = 0$.

25.2.

Although we got our equation (actually the infinite
system of difference-differential equations), the solution

is hard to find, and even if we get one it may be so complicated that it is practically useless. And besides this,
strange things may happen as we have seen in the case of
the birth process (in Section 24).

Therefore, it is of great practical interest to know if
there exists the equilibrium solution, that is the solution
which may be reached after a sufficiently long time. We
found such a solution in the queueing problem in Section 22
and also in the learning model in Section 20, but we have
seen that there is no such a solution for the birth process.

Thus, we shall assume that the following limit exists
for every j and is independent of the initial state i:

$$\lim_{t \to \infty} p_{ij}(t) = P(j)$$

and that P is a proper distribution:

$$\sum_j P(j) = 1.$$

It is intuitively clear that then $dp_{ij}(t)/dt \to 0$. Passing
to the limit with $t \to \infty$ in equation (*), one obtains the
so called equilibrium or steady-state equations:

$$
\begin{array}{ll}
-\lambda_0 P(0) + \mu_1 P(1) = 0 & j = 0 \\
\\
\lambda_{j-1} P(j-1) - (\lambda_j + \mu_j) P(j) + \mu_{j+1} P(j+1) = 0 & j > 0
\end{array}
$$

and if the number of states is finite, the last equation is:

$$\lambda_{n-1} P(n-1) - \mu_n P(n) = 0, \quad j = n.$$

These equations are very common in practical applications,
and may be solved explicitly. Proceeding as in Section 22
(by successive elimination) we can express the solution in
the form:

$$P(j) \; = \; P(0) \; \frac{\lambda_0 \lambda_1 \cdots \lambda_{j-1}}{\mu_1 \mu_2 \cdots \mu_j}$$

where $P(0)$ must be determined from the fact that all $P(j)$
add to 1. It is easy to check that solutions obtained in
Sections 20 and 22 can be easily obtained from this formula.
If you think that we pulled your leg by going through lengthy
calculations in these sections to obtain the result which we
now got effortlessly, rest assured that we have been far from
such evil thoughts. We simply chose this way to show you the
beauty of the mathematical models of general form. In this
vein we may add that the birth and death process is itself a
special case of the more general Markovian models.

Example 1: (Telephone exchange). The number of lines in a
telephone exchange is large, and may be assumed infinite.
Let Y_t be the number of busy lines at time t. Fluctuations
are caused by arrivals of calls (births) and by termination
of conversations (deaths). We shall assume that the input
is Poissonian and that deaths are proportional to the number
of busy lines. Translating this description into birth and
death rates, we have for all $0 \leq j < \infty$:

$$\lambda_j \; = \; \lambda, \qquad \mu_j \; = \; j\mu.$$

Equilibrium equations can be easily written down, and the solution is found to be:

$$P(j) = \frac{A^j}{j!} e^{-A}, \qquad \text{where } A = \lambda/\mu, \qquad j = 0,1,\ldots$$

again Poisson!

Example 2: (Servicing of machines). Suppose that in the work shop there are n machines and as usual they break down from time to time. Let Y_t be the number of nonworking machines. We shall assume that the probability of the breakage is proportional to the number of operating machines (births). The broken machine is put back into service and the time needed for repair is assumed to be the exponential life time. Thus, deaths correspond to the decrease in the number of inoperative machines. Again, translation into our birth and death coefficients yields:

$$\lambda_j = (n-j)\lambda, \qquad \mu_j = j\mu \qquad \text{for } j = 0,1,\ldots,n$$

(we have tacitly assumed independence of the machines).

Equilibrium equations can be easily written down, and the solution is found to be:

$$P(j) = \binom{n}{j} p^j (1-p)^{n-j}, \qquad \text{where } p = \lambda(\lambda+\mu)^{-1}, \quad j = 0,1,\ldots,n,$$

again binomial! (Compare this with subsection 21.2.)

25.3

In closing our discussion, it is hard to omit mentioning
how a trivial change leads to serious consequences. If you
look again at our balance equation in subsection 25.1, then
interchanging h with t does not affect the equation at
all. Yet, it leads to very important new results. Indeed,
h is now on the left side and this means that we now consider transitions before the instant t. Proceeding exactly
in the same manner as before, we obtain the dual system of
equations for $p_{ij}(t)$ -- we leave cheerfully their derivation
to the reader:

$$\frac{dp_{ij}}{dt}(t) = \lambda_i p_{i+1,j}(t) - (\lambda_i + \mu_i) p_{ij}(t) + \mu_i p_{i-1,j}(t)$$

with the same initial conditions as before. This new equation is called the backward equation (it has nothing wrong
with it -- the name signifies only the limit operation "on
the back"). Note that in the backward equation the second
index j is fixed, whereas in the forward equation it is
the first index i which is fixed. In the forward equation
the transitions are to the state j, but in the backward
equation the transitions are from the state i. It can be
shown that in most situations (in particular when the number of states is finite) both equations have the same solution. And let us leave worrying about different solutions
of these equations to Ph.D.'s in Mathematics.

Suppose that the limit P(j) exists. Then, passing
to the limit as t → ∞, the backward equation yields the
triviality 0 = 0. Do not, however jump to the conclusion
that the backward equation is useless. Just the opposite.
It is much more important than the forward equation. Indeed,
all problems concerning (continuous) life times in the
birth and death processes are described by the backward
equations. If you look closer at equations for the waiting
time distributions $G_n^c(t)$ in Section 23, you will notice
that we have there the backward equation (with $\lambda_i = 0$ and
$\mu_i = \mu$). Although we hinted at this situation from time to
time, we cannot enter into the fascinating study of the
backward equation and shall content ourselves with a rather
interesting example. See also the following Subsection
25.4 for more substantial application.

Example 3: (Linear growth). Suppose that intitially there
are i customers in the system, where i ≥ 1. Due to
fluctuations, it is possible that at some instant the number of customers will drop to 0. We would like to find the
probability of this extinction. There are many situations
which are covered by this model, the most famous being the
"extinction of family surnames." Another is the busy period
in the queueing system.

We shall consider the case of linear coefficients:

$$\lambda_i = i\lambda, \qquad \mu_i = i\mu, \qquad \text{for } i = 0,1,\ldots .$$

Clearly for $i = 0$, both coefficients vanish and it is impossible to leave the state 0 (thus, 0 is the absorbing state). We shall assume that $i > 0$, and shall consider the transition pr. from i to $j = 0$, namely $p_{i0}(t)$. It is intuitively clear that $p_{i0}(t)$ is in fact the distribution function of the (life) time to extinction. On the other hand, $p_{00}(t) \equiv 1$.

The backward equation for $j = 0$ and for $i > 0$ is:

$$\frac{dp_{i0}}{dt}(t) = i\lambda p_{i+1,0}(t) - i(\lambda+\mu)p_{i0}(t) + i\mu p_{i-1,0}(t).$$

Differentiation will show that the solution is of the form:

$$p_{i0}(t) = (\frac{\mu}{\lambda})^{i} (\frac{e^{(\lambda-\mu)t} - 1}{e^{(\lambda-\mu)t} - \frac{\mu}{\lambda}})^{i}, \quad i = 1,2,\ldots .$$

Passing to the limit with $t \to \infty$, it is easy to see that:

$$p_{i0}(\infty) = \begin{cases} 1 & \text{for } \lambda < \mu \\ \\ (\frac{\mu}{\lambda})^{i} & \text{for } \lambda \geq \mu \end{cases}.$$

Therefore, when $\lambda < \mu$, $p_{i0}(t)$ is a proper d.f. and the pr. that the time to extinction is finite is 1. Hence the extinction will take place.   On the other hand, when $\lambda > \mu$

then the pr. of extinction is less than one, and there is a

positive pr. $1 - p_{i0}(\infty)$ that the number of customers in

the system will increase over all bounds.   For $\lambda = \mu$,

extinction will take place, but the average time will be

infinite.   (Does this remind you about the fate of our rab-

bit in Section 18;   and about the explosion in Section 24 ?).

****

25.4

        Consider our birth and death process $(Y_t, \ 0 \leq t < \infty)$,

where Y_t represents the number of customers present at time

t.   Suppose that there is a critical level a, which may be

established, say, by management, with the following property.

If the number of customers is at least a, then the process

develops without any interruptions.   However, at the first

instant at which the number of customers drops below a, an

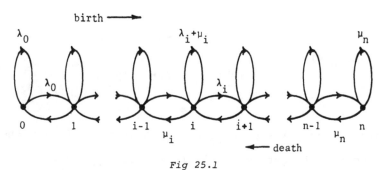

Fig 25.1

Birth and death process.

alarm sounds and the process stops abruptly. We will be interested in the probability of this happening, and we shall also look at the process before the alarm sounds.

Write $H = [0, 1, \ldots, a - 1]$ and $H^c = [a, a + 1, \ldots]$. We shall define the first entrance time T to a set H by

$$T = \inf(t: t > 0, Y_t \in H)$$

and $T = \infty$ when the process never enters H. We shall say in a rather morbid fashion that our process (Y_t) is killed at the first entrance time T to a set H. We also may call H a taboo set, as we are interested that our process continue its development safely. To study transitions avoiding set H, we shall introduce taboo pr.'s defined by

$$_H p_{ij}(t) = \text{pr}(Y_t = j, T > t | Y_o = i).$$

This is the transition pr. from state i to state j, avoiding taboo set H between them. Indeed, the first entrance to H must occur after time t.

It should be mentioned that we ignore here a serious problem of validity of the above expression. The reader must rest assured that this T is a well-defined life time (a stopping time, as a matter of fact), and that $_H p_{ij}(t)$ make sense. However, by definition, T is zero when the process starts in H. Thus, taboo pr.'s vanish when i or j are in H, so effectively $_H p_{ij}(t)$ are considered for i and j in H^c.

It is intuitively clear that taboo pr,'s satisfy the "balance equation" of the form

$$_Hp_{ij}(t + h) = \sum_{k \in H^c} {}_Hp_{ik}(h) \, {}_Hp_{kj}(t), \qquad t \geq 0, \ h \geq 0$$

with the same justification as in Subsection 25.1 (remember: H must be avoided).

Strictly speaking we are considering not the process itself, but rather its modification, which coincides with the Y process as long as it stays outside H, and for which all states in H may be considered absorbing. Indeed, we are interested in the first entrance time T to the set H—what happens afterward is of no concern to us. In other words, the original process Y is as good as dead once it has misfortune to visit H. It is therefore clear that the birth and death coefficients λ_i and μ_i for the modified process are the same as those for the original process provided $i \in H^c$; they all vanish when $i \in H$. Moreover, the structure of the birth and death process implies that in the short time interval, the set H may only be entered at state a - 1 by jump from state a.

Consequently, by the argument we used before, we obtain the backward equation for taboo probabilities $_Hp_{ij}(t)$, analogous to that stated in Subsection 25.3. Thus, for $j \in H^c$,

$$d_{H}p_{ij}(t)/dt = \lambda_i \, {}_Hp_{i+1,j}(t) - (\lambda_i + \mu_i){}_Hp_{ij}(t) + \mu_i \, {}_Hp_{i-1,j}(t)$$

for i > a, and for i = a,

$$d_{H}p_{aj}(t)/dt = -(\lambda_a + \mu_a){}_Hp_{aj}(t) + \lambda_a \, {}_Hp_{a+1,j}(t),$$

with the initial conditions

$$_H p_{ii}(0) = 1, \qquad _H p_{ij}(0) = 0, \qquad \text{for } i \neq j$$

(for both i and j in H^C).

Let $D_i(t)$ be the distribution function of the life time T, conditional on the initial state i. Evidently, the complementary distribution is

$$pr(T > t | Y_o = i) = D_i^C(t) = \sum_{j \in H^C} {}_H p_{ij}(t),$$

$$i \in H^C,$$

with $D_i^C(t) = 0$ for $i \in H$.

Hence, addition of equations for taboo pr.'s over all $j \in H^C$ yields the basic equation for $D_t^C(t)$ for $i \in H^C$:

$$d \; D_i^C(t)/dt = \mu_i D_{i-1}^C(t) - (\lambda_i + \mu_i) D_i^C(t) + \lambda_i D_{i+1}^C(t),$$

for $i > a$, and for $i = a$,

$$d \; D_a^C(t)/dt = -(\lambda_a + \mu_a) D_a^C(t) + \lambda_a D_{a+1}^C(t),$$

with initial condition $D_i^C(0) = 1$ for $i \in H^C$.

Unfortunately, these equations are difficult to solve, except in some special cases (one such case we have seen in Section 23, and the other is in Example 3 here).

It is therefore of practical importance to look for the limiting probability that T is finite, denoted $D_i = \lim_{t \to \infty} D_i(t)$, or for the complementary pr.:

$$D_i^C = pr(T = \infty | Y_o = i) = \lim_{t \to \infty} D_i^C(t).$$

(Warning: Do not exchange the indicated limit with the infinite sum over H^C, because $_H p_{ij}(t) \to 0$, but D_i^C need not

vanish.)   It is intuitively clear, however, that d $D_i^C(t)/$
dt $\rightarrow$ 0.   Passing to the limit with t $\rightarrow$ ∞ in the above equations, one obtains

$$0 = \begin{cases} \mu_i D_{i-1}^C - (\lambda_i + \mu_i)D_i^C + \lambda_i D_{i+1}^C, & i > a. \\ -(\lambda_a + \mu_a)D_a^C + \lambda_a D_{a+1}^C, & i = a. \end{cases}$$

Remembering that $D_i = 1$ for $i < a$, we can re-write the above
equations as a single equation for D_i:

$$\boxed{\mu_i D_{i-1} - (\lambda_i + \mu_i)D_i + \lambda_i D_{i+1} = 0, \qquad i \geq a}.$$

This is the fundamental equation for the absorption pr. D_i,
when starting at i.   This equation should not be confused
with a dual equation for P(j) in Subsection 25.2.

 We now solve this equation for D_i.   As the first step,
re-write the equation in the form

$$D_{i+1} - D_i = (\mu_i/\lambda_i)(D_i - D_{i-1}),$$

Hence, solving recursively one obtains

$$D_{i+1} - D_i = \gamma_i(D_a - 1),$$

where

$$\gamma_i = \frac{\mu_a \mu_{a+1} \cdots \mu_i}{\lambda_a \lambda_{a+1} \cdots \lambda_i},$$

Addition of these equations gives

$$D_{i+1} - D_a = (D_a - 1) \sum_{k=a}^{i} \gamma_k, \qquad i \geq a.$$

Passage to the limit with i $\rightarrow$ ∞, yields

$$D_\infty - D_a = (D_a - 1) \, \Gamma,$$

where

$$\Gamma = \sum_{k=a}^{\infty} \gamma_k.$$

It is intuitively clear that $D_\infty = 0$, which means that there is no chance to reach H from infinity. Hence, D_a is found to be

$$D_a = \Gamma/(1 + \Gamma).$$

Suppose that $\Gamma = \infty$. Then, $D_a = 1$, and it follows from the formula for D_{i+1} that all D_i are constants, equal to 1.

On the other hand, when $\Gamma < \infty$, then the same equation gives

$$D_i = (1 + \Gamma)^{-1} \sum_{k=i}^{\infty} \gamma_k, \qquad i \geq a.$$

This is the required explicit solution for probability D_i of reaching H from a state i in H^c.

Several interesting applications are obtained in the situation when the taboo set H consists of a single state O.

Example 4: Linear growth. This is the continuation of Example 3. Here $H = [O]$ and $a = 1$, so D_i represents the probability of extinction, when starting with i customers. Clearly,

$$\gamma_i = (\mu/\lambda)^i, \qquad i \geq 1$$

and $\Gamma = \infty$ if $\mu/\lambda \geq 1$, and $\Gamma < \infty$ if $\mu/\lambda < 1$. From the description of the process, it is clear that

$$D_i = p_{i0}(\infty).$$

For still another approach see Example 1 in Section 26.

Example 5: Busy period. Consider a simple queue discussed
in Sections 22 and 23. Take again H = [0] and a = 1 (single
server). The busy period T is the first entrance time to
zero, that is, the time when the number of customers drops
to zero for the first time. The proper busy period refers
to initial state i = 1. Then D_i is the pr. that (starting
with i customers) the busy period is finite. Recall that
the intensity has been defined as $\rho = \lambda/\mu$. Clearly,

$$\gamma_i = (1/\rho)^i \qquad i \geq 1$$

and $\Gamma = \infty$ if $\rho \leq 1$, and $\Gamma > \infty$ if $\rho > 1$. Hence,

$$D_i = \begin{cases} 1 & \text{if } \rho \leq 1. \\ (1/\rho)^i & \text{if } \rho > 1. \end{cases}$$

Observe that when $\rho < 1$, then there is the steady-state
solution and D_i is identically one, so the busy period will
be finite with pr. one. Its mean is known to be $i(\mu - \lambda)^{-1}$.
When $\rho > 1$, there is no steady-state solution, and probability that the busy period be infinite is positive, namely
$1 - D_i$. For $\rho = 1$, there is no steady-state solution, the
busy period is finite but its mean is infinite. For extension, see Example 4 in Section 26.

Example 6: Waiting time. Consider the waiting time discussed in Section 23. By a stretch of the imagination we

can convince ourselves that the waiting time is the first
entrance time to a set H = [0] with a = 1, but for the death
process with death coefficient $\mu_i = \mu$. In this case,
$D_i(t) = G_i(t)$, $i \geq 1$.

****

Thus, we return to where we started -- to talking
about life times: our main topic of discussion. And each
life begins and ends -- so it seems fit that we conclude
our investigations with a talk about birth and death.

****

Section 26: Branching Process

Consider a situation when initially there is a single object
and at some time later it splits into several objects of the
same type. Each of these newly formed objects splits again
as time progresses, producing further objects, and so on.
It may happen that an object just dies without producing
"offspring." Consequently, population size will fluctuate
with time; it may grow over all bounds, or may drop to zero
and become extinct; or perhaps it may somehow stabilize.

Such a system has numerous applications in the physical and social sciences. Typical examples include a colony
of bacteria that multiply by splitting, various chain reac-
tions involving splitting of particles (radiation or the
atomic bomb), the spread of epidemics and of rumors, "survival of family names" (counting sons only!), extinction of
species, flows over tree graphs (with branches), customers
arriving at a service facility (busy period), and so on.

The model is indeed quite general, and its name, the "branch-ing" (or multiplicative) process, reflects well its struc-ture.   Its mathematical description presents many interest-ing features, as we are going to see—never mind complexity of argument, as basic ideas are simple.

Let Y_t denote the population size at time t.   Clearly, Y_t assumes values 0, 1, 2, ...   For convenience of presentation, we shall say that Y_t represents the number of objects or individuals at time t (say, people, animals, or machines).   We shall also refer to objects obtained by the splitting procedure as descendents, offspring, or progeny and to initial individuals as ancestors.

The family $(Y_t, 0 \leq t < \infty)$ of random variables Y_t is called a branching process, and it describes the develop-ment with time of generations of objects initiated by ancestors.   This vague definition we shall replace by a more precise one.   Our primary task would be to find out when the population becomes extinct in a finite time and the probability of such an event.   This fascinating problem has received considerable attention in the literature, and clearly it is of great theoretical and practical interest.

26.1.

Let us now proceed to mathematical formulation—it will be analogous to that we have seen in this chapter, but not exactly the same.   We assume that (Y_t) form a Markovian process, and we shall consider the transition probabilities

$$p_{ij}(t) = pr(Y_t = j | Y_0 = i)$$

$$\text{for } i,j = 0, 1,\dots , t \geq 0,$$

that at time t the system is in a state j (population size is j), given that it was in state i initially. We shall impose an assumption of homogeneity, which means that this probability $p_{ij}(t)$ is the same for any pair Y_{t+s} and Y_s irrespective of s.

A Markov property implies that transition probabilities satisfy the "balance equation," which now has the form

$$p_{ij}(t + h) = \sum_{k=0}^{\infty} p_{ik}(t)p_{kj}(h), \qquad t \geq 0, h \geq 0.$$

This is the same equation already noted in Section 25.1 and admits of the same interpretation. We shall use it to derive equations for $p_{ij}(t)$. First, however, we must specify our process in a more precise way. The branching process is defined by the following properties:

A1: Each individual acts independently.

A2: Each individual has an exponential life time with mean $1/\gamma$.

A3: Each individual produces offspring with the same distribution b(k), $k \geq 0$, such that

$$0 < b(0) < 1, \qquad b(1) = 0, \qquad \sum_{k=0}^{\infty} b(k) = 1,$$

with mean m, assumed finite.

A4: The initial population is at least one, and if the size of population drops to zero, the process terminates.

   As in the prevous sections, we shall consider all
possible transitions that may take place during a time
interval (t,t + h) of a short length h > 0. Suppose that
at time t there are exactly i $\geq$ 1 individuals present—this
means that i exponential life times are in progress. We
assume that during subsequent duration h only one individual
can die; the pr. of several individuals dying simultaneously
is negligible.

   We found in Section 6 that the shortest life time
(among exponential ones) is also exponential, but with a
parameter iγ. Hence, by argument from Section 3, the
conditional pr. of termination of state i during (t,t + h)
is

$$1 - p_{ii}(h) = i\gamma h \qquad \text{(approximately).}$$

On dying, an individual splits into several new individuals,
resulting in a change of state from i to j, say, and this can
be achieved by producing j - i + 1 offspring. Thus, the
conditional pr. that during an interval (t,t + h) when
h > 0 is small, a transition to state j occurs, given that
at t the system was in state i, is (approximately)

$$p_{ij}(h) = i\gamma b(j - i + 1)h, \qquad j \geq i - 1, \ i \geq 1.$$

In particular, $p_{1j}(h) = \gamma b(j)h$ corresponds to the situation
described at the beginning of this section (with initial
population i = 1). Observe that if an individual produced
at death exactly one offspring, the size of the population
would not change. Imposing the requirement that death
should coincide with a change of state means that we must

have b(1) = 0. On the other hand, the assumption b(0) > 0

is essential because it takes care of a decrease in popula-

tion (and hence allows for eventual extinction):

$$p_{i,i-1}(h) = i\gamma b(0)h \qquad \text{(approximately)}, \quad i \geq 1.$$

Note that state 0 is absorbing, so $p_{00}(t) = 1$ for all t, and

can be reached during (t,t + h) only from state i = 1, with

pr. $\gamma b(0)h$, approximately. Clearly,

$$i\gamma \sum_{j=i-1}^{\infty} b(j - i + 1) = i\gamma, \qquad i \geq 1.$$

Although our branching process is Markovian, it is

not a birth and death process, as defined in Section 25

(where only transitions from state i to neighboring states

i - 1 and i + 1 were allowed during intervals of small

length h). Here we have transitions to any state $j \geq i - 1$

for $i \geq 1$. Obviously, no transitions are possible from

state i = 0.

Proceeding as in Section 25 (by starting with the

balance equation and considering both contingencies with t

and h interchanged), we look at the increment ratio

$$\frac{p_{ij}(t + h) - p_{ij}(h)}{h} \qquad \text{for } h > 0.$$

This equals

$$\sum_{k \neq j} p_{ik}(t) \, p_{kj}(h)/h - p_{ij}(t)(1 - p_{jj}(h))/h$$

for the "backward" argument. Substituting the expressions

for infinitesimal transition probabilities $p_{ij}(h)$ and passing to the limit with h → 0 (this step requires serious
mathematical justification), one obtains the forward and
backward equations for transition probabilities:

$$dp_{ij}(t)/dt = -p_{ij}(t)j\gamma + \sum_{k=1}^{j+1} p_{ik}(t)b(j - k + 1)k\gamma \quad (F)$$

$$dp_{ij}(t)/dt = -i\gamma p_{ij}(t) + i\gamma \sum_{k=i-1}^{\infty} b(k - i + 1)p_{kj}(t) \quad (B)$$

for i ≥ 1, j ≥ 0, and t ≥ 0, with the same initial condition

$$p_{ij}(0) = 0 \qquad \text{for } i \neq j \qquad \text{and} \qquad p_{ii}(0) = 1.$$

It is difficult to handle these frightening equations. The
best we can do is to convert them into something simpler,
but there is a price for doing that. As a preliminary step
in this direction, we need to take a closer look at the
structure of our branching process and to specify exactly
what is meant by extinction. Then, using equation (B), we
will be able to determine the probability of extinction.

26.2

     We shall now examine transition probabilities $p_{ij}(t)$
in some detail. Suppose that initially at t = 0 there are
just two individuals (i = 2). How can we have j individuals
later at time t? This is only possible if one individual
produces, say, k offspring and the other j - k. By assumption, individuals reproduce independently, so for the first

individual we have only transition l → k with probability $p_{1k}(t)$, and for the second one transition l → j - k with probability $p_{1,j-k}(t)$. Hence, the transition from i = 2 to j has (by independence) the probability given by convolution:

$$p_{2j}(t) = \sum_{k=0}^{j} p_{1k}(t)p_{1,j-k}(t) = \sum_{k+r=j} p_{1k}(t)p_{1r}(t)$$

(see the continuous analog in Section 5).

In the same fashion, we can repeat the argument for i = 3 individuals as ancestors. Hence, by induction, starting with i ancestors we obtain the basic formula

$$p_{ij}(t) = \sum p_{1k_1}(t)p_{1k_2}(t)\cdots p_{1k_i}(t), \qquad j \geq i - 1, \; i \geq 1$$

where the (i - 1)-tuple summation is extended over all k_1, k_2, ..., k_i such that

$$k_1 + k_2 + \cdots + k_i = j.$$

Indeed, the size of population at time t, starting with i individuals, is distributed as the sum of i independent populations starting with one individual. This fact—and the above formula—provides the principal characteristic of the branching process. We shall refer to it as to the branching property. Evidently this corresponds to the random sum of i.i.d. summands, as discussed in Section 15. With some imagination, we can express the size of population at time t + s in terms of size at time s by the relation

$$Y_{t+s} = N_1(t) + N_2(t) + \cdots + N_{Y_s}(t)$$

where each $N_v(t)$ is the number of offspring produced during
interval of length T by the v-th individual present at time
s (in particular, for s = 0).

****

All the above ideas can be put in a simpler form with
the help of generating functions. These are discrete ana-
logs of Laplace transforms and are merely special cost
functions introduced in Problem 28, Chapter 2 (see also
Section 23). Thus, define

$$h(z) = \sum_{k=0}^{\infty} b(k)z^k \qquad \text{for } |z| \le 1$$

with $h(0) = b(0)$, $h(1) = 1$, and

$$\phi_i(t,z) = \sum_{j=0}^{\infty} p_{ij}(t)z^j \qquad \text{for } |z| \le 1$$

with

$$\phi_i(0,z) = z^i, \qquad \phi_i(t,0) = p_{io}(t), \qquad \phi_i(t,1) = 1$$

for $i \ge 1$, and $\phi_o(t,z) = 1$ for $i = 0$.

Multiplying the above branching property expression
for $p_{ij}(t)$ by z^j and adding, we have immediately (by convolution property) that

$$\phi_i(t,z) = [\phi(t,z)]^i, \qquad i \ge 0, \qquad (*)$$

where we write $\phi = \phi_1$. In this form the basic branching
property is easier to handle; see comments in Problem 28,
Chapter 2.

****

As the first application, we have the following iterative
property:

$$\phi(t + s,z) = \phi[t,\phi(s,z)].$$

This follows from the balance equation

$$[\phi(t,z)]^i = \sum_{j=o}^{\infty} \sum_{k=o}^{\infty} P_{ik}(t)\, P_{kj}(s)\; z^j$$

$$= \sum_{k=o}^{\infty} P_{ik}(t) \sum_{j=o}^{\infty} P_{kj}(s)z^j$$

$$= \sum_{k=o}^{\infty} P_{ik}(t)\, [\phi(s,z)]^k$$

$$= [\phi(t,\psi(u,u))]^i$$

****

We can now simplify our equations for transition
pr.'s. Multiply both sides of equations (F) and (B) by z^j,
and add over all j. These equations are then converted into
partial differential equations of the form (see Problem 30
for details)

$$\partial\phi_i/\partial t = \gamma[h(z) - z]\, \partial\phi_i/\partial z, \qquad i \geq 1, \qquad \text{(F)}$$

$$\partial\phi/\partial t = \gamma\,[h(\phi) - \phi], \qquad\qquad\qquad \text{(B)}$$

where we wrote $\phi_i = \phi_i(t,z)$. These equations are easier to
handle and may be solved explicitly at least in some interesting cases.

****

It is easy to check (see Problem 31) that the conditional expectation of Y_t, given the initial number i of ancestors, has the form

$$\mathbb{E}\,(Y_t|Y_0 = i) = i\mathbb{E}\,(Y_t|Y_0 = 1) = ie^{\gamma(m-1)t},$$

$$t \geq 0,\ i \geq 1$$

where m is the average size of progeny. This is a rather interesting result. The average size will increase to infinity with time if $m > 1$ and will decrease to zero when $m < 1$. For $m = 1$, it will stay constant—this, however, may be misleading because it does not provide full information about behavior for large t.

26.3

We are now ready to attack the main problem of extinction. We shall find out that in our branching process only extinction or explosion is possible.

Suppose that we start with $i \geq 1$ ancestors at time $= 0$. Let T denote the time needed for population size to drop to zero; as state 0 is absorbing, the process (Y_t) is killed when it enters state 0. In other words, extinction takes place. Thus, T is the time to extinction defined as the time to first entrance to absorbing state 0:

$$T = \inf(t:\ Y_t = 0)$$

and $T = \infty$ when the process never enters 0. Denote the distribution of T, conditional on the initial state i, by

$$D_i(t) = pr(T \leq t|Y_0 = i), \qquad i \geq 1,\ t \geq 0.$$

Now, the event $(T > t)$—that the first entrance to 0 occurs after time t—is the same that the process stays away from 0 up to time t. Indeed, starting with i individuals, when $i \geq 1$, the process avoids state 0 in its fluctuations. We express this by saying that transition probabilities from state i to state j (when both i and j are different from zero) are <u>taboo probabilities</u> (i.e., avoiding taboo 0):

$$P_{ij}(t) = pr(Y_t = j, t < T | Y_0 = i), \qquad i, j \geq 1.$$

Hence

$$pr(t < T | Y_0 = i) = \sum_{j=1}^{\infty} P_{ij}(t) = 1 - P_{io}(t).$$

This gives us the required distribution

$$D_i(t) = P_{io}(t) = [P_{1o}(t)]^i = D_1^i(t)$$

where the last expression follows from the branching property (*) evaluated at $z = 0$. So it is sufficient to find only $P_{1o}(t) = D_1(t) = \phi(t,0)$. The backward equation (B) is handy here:

$$dD_1(t)/dt = -\gamma D_1(t) + \gamma \sum_{k=o}^{\infty} b(k) D_1^k(t)$$

$$= \gamma(h(D_1(t)) - D_1(t)) \qquad (**)$$

with the initial condition $D_1(0) = 0$.

We now define the probability of extinction as the probability that population size drops to zero in finite time when starting with one individual:

$$\lim_{t \to \infty} P_{1o}(t) = pr(T < \infty | Y_0 = 1) = D_1.$$

Clearly, the probability of extinction when starting with i individuals is

$$D_i = D_1^i.$$

To solve our problem we need to find D_1 only. This leads us to the famous <u>branching theorem</u>, which asserts that the extinction probability D_1 is the smallest non-negative root of the equation

$$h(z) = z, \qquad\qquad (***)$$

and

$$D_1 = 1 \qquad \text{iff} \qquad m \le 1, \qquad \text{or equivalently,}$$

$$D_1 < 1 \qquad \text{iff} \qquad m > 1.$$

Thus, if the average size of progeny is strictly larger than 1, then there is a positive probability $1 - D_1$ that the population will increase to infinity in finite time (explosion). In the opposite case $m \le 1$, the population will die out in finite time. This famous result, although intuitively obvious, nevertheless required rather heavy machinery to confirm it. Well, things become more complicated toward the end of the book.

Let us verify the branching theorem. It is intuitively clear that as $D_1(t)$ tends to a constant D_1, its derivative would tend to zero. Passing to the limit with $t \to \infty$ in the backward equation (**) for $D_1(t)$, one obtains

$$0 = h(D_1) - D_1,$$

So D_1 satisfies equation (***). How many solutions may

this equation have? Clearly, $z = 1$ is a solution, but $z = 0$ is not because $b(0) > 0$. Plot graphs of functions $y = h(z)$ and $y = z$ for the region $0 \le z \le 1$. Their intersections represent the solutions of equation $h(z) = z$. Note that both functions $h(z)$ and z are continous and increasing; $h(z)$ is a power series and z is the diagonal line. Compare their slope at $z = 1$, namely, $h'(1) = m$ and 1. It is now clear that if the slope of $h(z)$ at $z = 1$ is strictly larger than the slope of a diagonal line, i.e., $m > 1$, then the curve $h(z)$ will cross the diagonal z at exactly one point $z_0 < 1$. This z_0 is our D_1, strictly less than 1. In the opposite case, i.e., $m \le 1$, the curve $h(z)$ will lie entirely above the diagonal z, with $z_0 = 1$ the only point of intersection. Thus, $D_1 = 1$ in this case. See Figure 26.1.

$$*************$$

Consider now the average time to extinction (when the initial population is $i \ge 1$), and write

$$M_i = \mathbb{E}(T \mid Y_0 = i).$$

Evidently, M_i is infinite when $D_i < 1$ (because $T = \infty$ with positive probability). Suppose therefore that $D_i = 1$, so $D_i(t)$ is the proper distribution function of a life time T and T is finite, so

$$M_i = \int_0^\infty D_i^c(t)dt.$$

Evaluation of M_i is rather complex, but fortunately we can draw important conclusions without the explicit form of M_i.

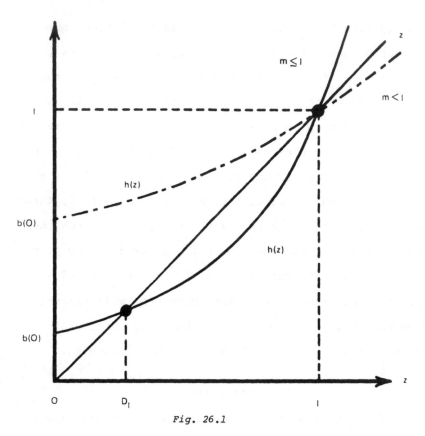

Fig. 26.1

Solution of Equation h(z) = z.

First, observe that the branching property $D_i(t) =$ $D_1^i(t)$ implies that T has the same distribution as the largest life time

$$W_i = \max(T_1, \ldots, T_i)$$

among independent identically distributed life times $T_1, \ldots,$ T_i, each with the same distribution $D_1(t)$. This is evident from our discussion in Section 6.

We must now attach a name to a discrete life time representing progeny, so let K be a random variable with distribution b and mean $\mathbb{E} K = m$. As $D_1 = 1$, so by the branching theorem we have $m \leq 1$.

Next, we may observe that $h(D_1(t))$ is the distribution function of a compound r.v.

$$W_K = \max(T_1, \ldots, T_K),$$

which represents the largest life time from a collection containing a random number K of i.i.d. life times. Indeed,

$$\text{pr}(W_K \leq t) = \sum_{k=0}^{\infty} \text{pr}(W_K \leq t | K = k) \text{pr}(K = k)$$

$$= \sum_{k=0}^{\infty} D_k(t) b(k) = h(D_1(t))$$

(see Section 15 for the similar argument).

We now re-write the backward equation (**) in the form

$$dD_1(t)/dt = \gamma(D_1^c(t) - \text{pr}(W_K > t)).$$

Integration from 0 to ∞ gives the relation

$$1 = \gamma[M_1 - \mathbb{E} W_K]$$

which would be helpful if $\mathbb{E} W_K$ were known. Unfortunately, this is not the case, and we must consider some approximations. Indeed, for z close to 1, we have approximately for $m \leq 1$,

$$h(z) - z \approx (1 - m)(1 - z)$$

so with $z = D_1(t)$ this corresponds to large t. Substitution

of the above formula into the equation (**) yields, after
integration.

$$1 \approx \gamma(1 - m) \int_0^\infty D_1^c(t) \, dt = \gamma(1 - m) \, M_1.$$

Hence for $m = 1$, we have $M_1 = \infty$ (see Example 1 below).

    <u>Warning</u>. It may be tempting to claim that $\mathbb{E} \, W_K$
equals $m M_1 = \mathbb{E} K \cdot \mathbb{E} T_1$ (by analogy with expressions that we
encountered earlier, say, in Sections 15 and 21, and Prob-
lem 29 in Chapter 2). This would be too simple to be true.
It would even mean that M_i are given by $i/\gamma(1 - m)$, which
is clearly contradicted by Example 1 (see, however, Example
4).

****

    This completes our discussion, and we are now ready
to look at some examples.

26.4.

    The examples given here exhibit interesting features,
despite that the calculations are rather involved (but
strongly recommended!).

<u>Example 1</u>: Linear process. Take $\gamma = \lambda + \mu$ and define

$$b(0) = \mu/\gamma, \qquad b(2) = \lambda/\gamma \qquad \text{and}$$

$$b(k) = 0 \text{ otherwise.}$$

It is clear that this choice gives the linear birth and
death process, discussed in Example 3 in Section 25, where
we found expressions for $p_{io}(t)$ and for D_i. Indeed, now

$$\gamma h(z) = \mu + \lambda z^2$$

so the solution of $h(z) = z$ yields

$$D_1 = \begin{cases} 1 & \text{iff} \quad \lambda/\mu \leq 1 \\ \mu/\lambda & \text{iff} \quad \lambda/\mu > 1 \end{cases}$$

in agreement with what we already found. Moreover, direct calculation yields

$$M_1 = (-1/\lambda) \log (1 - \lambda/\mu) \qquad \text{for } \lambda/\mu \leq 1.$$

Example 2: Let N be a discrete life time (assuming non-negative integers) with some distribution $p(k) = \text{pr}(N = k)$ for $k = 0, 1, 2,\ldots$. Define

$$b(k) = \text{pr}(N = k | N \neq 1) = p(k)/[1 - p(1)]$$
$$\text{for } k \neq 1 \text{ and } b(1) = 0.$$

With the generating function

$$P(z) = \sum_{k=0}^{\infty} p(k)z^k$$

it is easy to see that the equation $P(z) = z$ is equivalent to the equation $h(z) = z$.

Specialize now to the geometric distribution of the form

$$p(k) = (1 - \rho)\rho^k, \qquad \text{where } 0 < \rho < 1,$$

with mean $\rho/(1 - \rho)$ and

$$P(z) = (1 - \rho)/(1 - \rho z).$$

This gives

$$b(k) = (1 - \rho)\rho^k/[1 - (1 - \rho)\rho], \qquad \text{for } k \neq 1,$$

and easy calculation shows that the mean is

$$m = \frac{\rho}{1 - \rho} \frac{1 - (1 - \rho)^2}{1 - (1 - \rho)\rho}.$$

Thus, $h(z) = z$ reduces to the quadratic equation

$$\rho z^2 - z + 1 - \rho = 0$$

which has two solutions $(1 - \rho)/\rho$ and 1. Hence $m > 1$ is
equivalent to condition $\rho > 1/2$, so the extinction probability is

$$D_1 = (1 - \rho)/\rho < 1.$$

Example 3: Strange distribution. Take for the generating
function of progeny

$$h(z) = 1 - (1 - z)^a + a(1 - z)$$

where a is fixed such that $0 < a < 1$. It follows that $b(0) =$
a and $b(1) = 0$, but this distribution has infinite mean (m =
∞).

It can be verified by tedious solution of equation
(B) (See Problem 34) that

$$\phi(t,1) < 1 \qquad \text{for all } 0 < t < \infty.$$

Hence again we have the sub-stochastic case (probabilities
do not add to 1); see Section 24 for the similar situation.

Example 4: Busy period. Connections between a branching
process and a busy period in a queueing system are well
known and have received considerable attention in the

literature. We briefly remarked about such a possibility
in Section 25. Here we shall look at a different, and less
known, type of connection, which is interesting and rather
puzzling. We shall present basic ideas, without entering
into technical details.

Consider a simple queue discussed in Sections 22 and
23 but modified in the following manner. We still have a
single server, but customers now arrive in groups (batches)
of variable size. This is the group input, the groups
following Poisson input with rate λ, and their size is
distributed according to a specified distribution p (the
same for each group); p(k) is the pr. of having a group of
size k, so

$$\sum_{k=1}^{\infty} p(k) = 1$$

Departures from the system occur individually, according to
exponential service time with mean $1/\mu$.

Typical examples are provided by buses bringing
tourists, by letters found in your mailbox, by orders to be
processed, by power outages that must be taken care of, and
by many situations of this sort.

Let N_t represent the number of customers in the
system at time t (including the one being served). When
N_t = i, the following transitions are possible during the
short interval (t,t + h) with h > 0:

$i \rightarrow j$:   arrival of $j - i \geq 1$ customers,
with intensity $\lambda p(j - i)$, $i \geq 0$.

$i \rightarrow i - 1$: departure of a single customer, with

intensity μ, $i \geq 1$.

$i \rightarrow i$:   no change, with intensity $\lambda + \mu$ for

$i \geq 1$, and λ for $i = 0$.

The busy period for the process (N_t) is defined as the first entrance time T^* to state 0:

$$T^* = \inf \, (t: N_t = 0)$$

with $T^* = \infty$ when the process never reaches 0. Let $D_i^*(t)$ be the d.f. of T^*, conditional on initial state $N_0 = i$. The proper busy period refers to $i = 1$. Equations for $D_i^*(t)$ can be written (regarding state 0 as an absorbing state), but we shall not need them for our brief discussion (these are the backward equations, like those in Section 25).

Consider now the branching process (Y_t) for which we select the progeny distribution of the form

$$\gamma b(k) = \lambda p(k - 1), \qquad k \geq 2$$

$b(1) = 0$ and $\gamma b(0) = \mu$, where $\gamma = \lambda + \mu$. There is a curious relationship between these two processes (N_t) and (Y_t). Let A_∞ be the total number of individuals present throughout all time of the branching process. We may represent this by an integral

$$A_\infty = \int_0^\infty Y_t \, dt.$$

It turns out—believe it or not—that A_∞ has the same distribution as the busy period T^*, both conditional on initial state i:

$$\text{pr } (A_\infty \leq t | Y_0 = i) = D_i^*(t),$$

$$0 \leq t < \infty, \; i \geq 1.$$

In particular, $D_i^* = \lim_{t \to \infty} D_i^*(t)$ is the pr. that the busy period (starting with i customers) or equivalently that the total number A_∞ in the branching process, are both finite.

Suppose that $D_i^* = 1$. Using the expression for the conditional expectation $\mathbb{E}(Y_t | Y_0 = i)$, we can evaluate the mean of the busy period:

$$\mathbb{E} (A_\infty | Y_0 = i) = \mathbb{E} (\hat{T}^* | N_0 = i) = \frac{i}{\gamma(1 - m)}$$

where m is the mean of progeny distribution. Note that the average busy period is finite when m < 1 but it is infinite when m = 1. On the other hand, when $D_i^* < 1$, the busy perios T^* is infinite (this corresponds to overload of the system).

Perhaps it is worthwhile to repeat that in the present case extinction of the branching process (A_∞ finite) corresponds to the termination of the busy period T^*. Hence, the branching theorem referring to D_i (hence to m) also refers to D_i^*.

Problems: Chapter 4

1.   If customers arrive at the rate of 2 per minute, what is the pr. that the number of customers arriving in 2 minutes is:

(i)   exactly 3,   (ii)   3 or less,   (iii)   3 or more,

(iv)   more than 3,   (v)   less than 3.

Suppose there was initially 4 customers present.
What was the average number of customers present at the
end of a 2 minute period?

2.   Let $p_{ij}(t)$ be the Poisson distribution from Section
19. Show that

   (i) $\displaystyle\int_0^\infty e^{-\alpha t} p_{ij}(t)\ dt\ =\ \frac{\lambda^{j-i}}{(\lambda+\alpha)^{j-i+1}}$ for $j \geq i$;

   (ii)   the value of $dp_{ij}(t)/dt$ at $t = 0$ is $-\lambda$ if
         $i = j$, λ if $j = i + 1$ and 0 otherwise;

   (iii)   for fixed i and j, plot the graph of $p_{ij}(t)$
         as a function of t (see problems 4 and 5, Chapt. 3).

3.   Suppose that the random time T   (of Section 19) has
     density of the form:

$$f(t)\ =\ \frac{(\mu t)^{n-1}}{(n-1)!}\ e^{-\mu t}\mu,\qquad t > 0,\quad n = 1,2,\ldots\ .$$

Show that

$$\text{pr}(N_T = j)\ =\ \binom{n+j-1}{j}\frac{\mu^n\lambda^j}{(\mu+\lambda)^{n+j}},\qquad j = 0,1,\ldots$$

and $EN_T = n\lambda/\mu$.

4. For the Poisson input (Section 19), let φ be the cost function associated with states of the form:

$$\varphi(j) = z^j \qquad \text{for} \quad 0 < z < 1.$$

Show that the average conditional cost is

$$E(\varphi(Y_t) \mid Y_0 = i) = z^i e^{-\lambda t(1-z)} \leq z^i.$$

5. Calculate all pr.'s in Section 20 for the following conditions:

   (i) $a = b$, $\pi_0 \neq \pi_1$;

   (ii) $\pi_0 b = \pi_1 a$;

   (iii) $\pi_0 = \pi_1$, $a \neq b$.

   Plot the corresponding graphs for selected values at a, b, and π_0, π_1.

6. Let $\varphi(i)$ be the cost function $(i = 0,1)$ associated with the learning model in Section 20. Evaluate the average cost:

$$E\varphi(Y_t) = \varphi(0)\pi_0(t) + \varphi(1)\pi_1(t).$$

Show that the average cost is:

   (i) constant if $\varphi(0) = \varphi(1)$;

   (ii) $C(\pi_0 b - a\pi_1)e^{-(a+b)t}$ if $\varphi(0) = bC$, $\varphi(1) = -aC$

   (where C is constant).

7. Suppose that a clever and hungry monkey tries to reach a
    banana with a stick. Let $a/b = k > 1$. Show that the
    pr. that the monkey will eventually learn to catch the
    banana is

$$\frac{1}{2} < \frac{k}{1+k} < 1.$$

If the monkey receives a reward $c(k+1)$ for a correct
response, and $-c(k+1)$ for a wrong response, show that
the average reward at time t is

$$c[k-1 + 2(\pi_0 - \pi_1 k)e^{-(1+k)bt}].$$

8. Verify that $p_{ij}(t)$ from Section 20 satisfy the
    integral equations in Section 21, by

    (i)   direct substitution;

    (ii) comparison of Laplace transform

$$u_{ij}(\alpha) = \int_0^\infty e^{-\alpha t} p_{ij}(t)\ dt.$$

In both cases the Laplace transforms are:

$$u_{00}(\alpha) = \frac{\alpha + a}{\alpha(\alpha + a + b)}, \quad u_{01}(\alpha) = \frac{b}{\alpha(\alpha + a + b)}$$

$$u_{10}(\alpha) = \frac{a}{\alpha(\alpha + a + b)}, \quad u_{11}(\alpha) = \frac{\alpha + b}{\alpha(\alpha + a + b)}.$$

9. Differentiate expression $p_{ij}(t)$ from Section 21.2 and
    verify that it satisfies the birth and death equation
    with coefficients

$$\lambda_j = (n-j)b, \quad \mu_j = ja \qquad (j = 0,1,\ldots,n)$$

(see Example 2, in Section 25).

10. In the group of n lines (Section 21.2) suppose that there are j lines busy and n-j free. Let X_0 and X_1 be the idle period and the busy period for each line, where X_0 and X_1 have exponential distribution with parameters b and a, respectively.

    Define the function T of the state j to be the minimum of j life times X_1 and (n-j) life times X_0. Show that T has an exponential distribution with the parameter q_j = (n-j)b + ja (use $G_-^c(t)$ from Section 6).

11. In the queueing model (Section 22), let the average interarrival time be 10 minutes, and the average time spent at the server be 5 minutes. Show that

$$P(j) = (\tfrac{1}{2})^{j+1}, \quad j = 0,1,\ldots,$$

    and $E(Y_t)$ = 1. Show also that the average waiting time (Section 23) is E(W) = 5 minutes.

12. Suppose that the cost function associated with the steady state distribution in the queueing model (Section 22) is given by $\varphi(j) = z^j$, $|z| \le 1$. Show that the average cost is then $(1-\rho)/(1-\rho z)$.

13. Plot the graph of the mean number of customers in the system $\rho/(1-\rho)$, as a function of ρ (for $0 < \rho < 1$).

Interpret the limits when $\rho \to 0$ and $\rho \to 1$. Show that the mean is less than 1 when $\rho < \frac{1}{2}$; interpret.

14. For the waiting time W (Section 23) evaluate

$$EW^2 = 2 \int_0^\infty t W^c(t) \, dt,$$

and deduce that

$$\text{var } W = \frac{2\rho - \rho^2}{\mu^2(1-\rho)^2}.$$

15. For the waiting time W (Section 23) evaluate

$$\int_0^\infty e^{-\alpha t} W^c(t) \, dt,$$

and deduce (from Section 7) that the Laplace transform of the waiting time distribution is

$$\int_0^\infty e^{-\alpha t} \, dW(t) = (1-\rho) \frac{\mu+\alpha}{\mu-\lambda+\alpha}.$$

Derive again expressions for $E(W)$ and $\text{var}(W)$.

16. (i) In Section 23, verify that $W^c(t)/E(W)$ defines an exponential density with mean $(\mu - \lambda)^{-1}$ (Cf. Section 4). Interpret!

   (ii) Let V be the life time with this density. Show that

$$\mathbb{E} Y_t = \lambda \mathbb{E} V.$$

   This is a special case of the famous "Little formula" of rather wide applicability.

17. Show that the pr. of waiting more than the average waiting time is $\rho e^{-\rho}$.

18. Suppose that the cost function in the waiting time problem (Section 23) is linear $\varphi(t) = at + b$ (for $t \geq 0$). Show that the average cost of waiting is

$$E_\varphi(W) \quad = \quad b + \frac{\rho a}{\mu(1-\rho)} \ .$$

19. In the waiting time problem (Section 23), suppose that the pr. of no waiting is given, say p, and that the average waiting time is known, say m. Show that:

$$\lambda = \frac{(1-p)^2}{mp} \ , \qquad \mu = \frac{1-p}{mp} \ .$$

20. In the birth process (Section 24), let the transition pr. be given for $i \geq 1$ by:

$$p_{ij}(t) \quad = \quad \binom{j-1}{j-i}(1-e^{-\lambda t})^{j-i}e^{-\lambda it}, \qquad j \geq i.$$

By differentiation, verify that $p_{ij}(t)$ satisfies the backward equation of the form

$$\frac{dp_{ij}}{dt}(t) \quad = \quad \begin{cases} \lambda_i p_{i+1,j}(t) - \lambda_i p_{ij}(t) & \text{for } i \leq j-1 \\[2mm] -\lambda_i p_{ii}(t) & \text{for } i = j \end{cases}$$

with $\lambda_i = i\lambda$. By rearrangement of terms verify that the forward equation holds, too.

21. Suppose that the input stops altogether when the number
     of births reaches a fixed number, say n (this is
     called "finite input"), so

$$\lambda_j \;=\; (n-j)\lambda \quad \text{for} \quad j = 0,1,\ldots,n.$$

(i) Verify that

$$p_{ij}(t) \;=\; \begin{cases} \binom{n-i}{j-i} e^{-(n-j)\lambda t}(1-e^{-\lambda t})^{j-i}, & i \le j \le n \\[2ex] 0, & i > j \end{cases}$$

is the solution of the birth equation with the
above coefficient.

(ii) Show that $E(Y_t \mid Y_0 = i) = (n-i)(1-e^{-\lambda t})$,
      $\operatorname{var}(Y_t \mid Y_0 = i) = (n-i)(1-e^{-\lambda t})e^{-\lambda t}$.

22. Verify Examples 1 and 2 in Section 25.

23. Give the full derivation of the backward equation in
     Section 25.

24. With reference to Section 25, verify that the derivatives
     at t = 0 are:

$$\left.\frac{dp_{i,i+1}}{dt}(t)\right|_{t=0} = \lambda_i, \qquad \frac{dp_{i,i-1}}{dt}(t) = \mu_i,$$

$$\left.\frac{dp_{ii}}{dt}(t)\right|_{t=0} = -(\lambda_i + \mu_i).$$

25.   (Death process).   Show that the binomial distribution

$$p_{ij}(t) = \begin{cases} \binom{i}{j} e^{-\mu j t} (1-e^{-\mu t})^{i-j} & \text{for } 0 \le j \le i \\[2ex] 0, & \text{for } j > i \end{cases}$$

satisfies the death equation, i.e., $\mu_j = j\mu$ for

$j = 1,2,\ldots,i$, and $\lambda_j \equiv 0$. In this problem, one

starts with i items initially, and only terminations

are allowed, the state $j = 0$ being the graveyard.

26.   (Infinite number of lines).   This is a generalization

of the case presented in Section 21.2 when n is

infinite.   Consider the transition pr. of the form:

$$p_{ij}(t) = \sum_k r_i(j-k,t)\, q_i(k,t)$$

with summation over $\min(j-i,0) \le k \le j$, where

$$r_i(j-k,t) = \binom{i}{j-k} e^{-\mu t(j-k)} (1-e^{-\mu t})^{i-j+k}$$

and

$$q_i(k,t) = \frac{[A(t)]^k}{k!} e^{-A(t)}$$

with

$$A(t) = \frac{\lambda}{\mu}(1-e^{-\mu t}).$$

This form follows from that in Section 21.2 by passage

to the limit, as in Section 14.   Verify that $p_{ij}(t)$

satisfies the birth and death equations with

$$\lambda_j = \lambda, \quad \mu_j = j\mu \quad \text{for} \quad 0 \le j < \infty.$$

Check also that the steady state solution $P(j)$ exists (as $t \to \infty$) and coincides with that given in Example 1 (Section 25).

27. Let $p_{ij}(t)$ be the transition probability of the Poisson process (Section 19).

Show that:

$$p_{ij}(t) \to 0 \quad \text{as} \quad t \to \infty$$

but (summing over all even j) one has for $t \to \infty$:

$$\sum_{j \text{ even}} p_{ij}(t) = \begin{cases} e^{-\lambda t} \cosh \lambda t \to 1/2, & \text{when } i \text{ even} \\ e^{-\lambda t} \sinh \lambda t \to 1/2, & \text{when } i \text{ odd} \end{cases}$$

28. Let M be a random variable with d.f. G_∞ obtained in Problem 19, Chapt. 3. Consider a life time M^+ obtained by truncation of M at zero:

$$M^+ = \max(0, M)$$

(see Section 6).

Show that the d.f. of M^+ coincides with the d.f. W of the waiting time obtained in Section 23. (This constitutes another approach to the study of waiting times.)

29. Consider a birth process discussed in Section 24. Let T_k be the first entrance time to a fixed state k:

$$T_k = \inf(t: t > 0, Y_t = k)$$

and write for its d.f. conditional on initial state i < k:

$$\text{pr}(T_k \leq t | Y_o = i) = F_{ik}(t),$$

with density $f_{ik}(t)$ and Laplace transform $f^*_{ik}(\alpha)$.

Show that for i < k,

$$\alpha u^*_{ik}(\alpha) = f^*_{ik}(\alpha) = \frac{\lambda_i}{\alpha + \lambda_i} \cdots \frac{\lambda_{k-1}}{\alpha + \lambda_{k-1}} \to 1$$

as $\alpha \to 0$.

Hint: you may consider k as absorbing ($\lambda_k = 0$).

30. Derive the forward and the backward equations in transformed form, stated in Subsection 26.2.

31. Show that for the branching process,

$$\mathbb{E}(Y_t | Y_o = i) = ie^{\gamma(m-1)t}$$

and deduce that

$$\mathbb{E}\left(e^{-\gamma(m-1)(t+s)} Y_{t+s} | Y_s\right) = Y_s e^{-\gamma(m-1)s}.$$

32. Evaluate M_1 in Example 1 in Section 26. Try to find variance.

33. Complete the details of Example 2 in Section 26.

34. Consider the generating function from Example 3 in Section 26:

$$h(z) = 1 - (1 - z)^a + a(1 - z),$$

$$0 < a < 1.$$

(i)   Find the solution of the equation $h(z) = z$.

(ii) Show that $m = \infty$.

(iii) Find the explicit expression for probabilities $b(k)$.

(iv) Write the backward equation from Section 26 in the form

$$\int_z^\phi \frac{dx}{h(x) - x} = \gamma t, \qquad \text{where } \phi = \phi(t,z),$$

and show (evaluating the integral) that

$$(1 + a)(1 - \phi)^{1-a} = 1 - [1 - (1 - z)^{1-a}(1 + a)]$$
$$\times e^{-(1-a^2)\gamma t}.$$

Deduce that $\phi(t,1) < 1$.

35. A car salesman averages seven sales per week.

(i)   What is the average number of sales per day?

(ii) Assuming exponential distribution between consecutive sales, find the probability of having no sale at all during a week.

(iii) Suppose that the salesman is to be fired at the (end of the) first week that produces no sales. What is the probability that the salesman will be fired after 1 month on the job? (Assume independence of consecutive weeks.)

(iv) Verify that numerically probabilities in ii and

iii are close to each other and that both are very small.

36. Let $p_{ij}(t)$ be the transition pr. for the Poisson input Y_t with a parameter λ.

    (a)   Show that $p_{ij}(t)\lambda$ is the density of the time needed to move from state i to state j.

    (b)   Let $\phi(j) = aj + b$ be the cost function (where a and b are positive constants). Show that the average cost (conditional on i) is

$$\mathbb{E}\,(\phi(Y_t)\,|\,Y_0 = i) = a(\lambda t + i) + b.$$

37.   (a)   Suppose that in the simple queue the average inter-arrival time is 15 minutes and the average service time is 10 minutes. Find:

    (i)   The pr. that the waiting time will exceed half-an-hour

    (ii)   The pr. of no waiting at all

    (iii)   The average waiting time

    (b)   Suppose that the cost of operation in the simple queue is equal to a constant a when there is no waiting but increases proportionally with the waiting time when waiting time is positive. The corresponding cost function is given by

$$\phi(0) = a \qquad \text{when } t = 0$$
$$\phi(t) = bt \qquad \text{when } t > 0$$

where a and b are positive constants.

(i)   Find the average cost of waiting $\mathbb{E}\phi(W)$.

(ii) Suppose that the ratio a/b equals the average
      service time. Verify that the average cost in
      i must be larger than a.

38. Let Y_t be the number of customers in the simple queue
    under equilibrium conditions. Let W be the waiting
    time of a customer, and let ρ be the intensity.

    (i)   Show that $\mathrm{pr}(Y_t \geq k) = \rho^k$ for each k = 0, 1,
          ... .

    (ii) Show that $\mathrm{pr}(W) > 1/\mu)$ is an increasing func-
          tion of ρ (here $1/\mu$ is the average service
          time), and sketch its graph.

    (iii) Suppose that when Y_t is less than k, it costs
          nothing to operate the system. If, however,
          Y_t is k or more, then the cost of operation
          is constant and equal to c(k), depending on k.

          Find this constant c(k) when it is required
          that the average cost of operation should not
          depend on k.

39. Suppose that the joint density f of two life times X
    and Y has the following form:

$$f(x,y) = 6x/a^3 \qquad \text{for } (x,y) \text{ in T}, \qquad \text{and}$$
$$f(x,y) = 0 \qquad \text{otherwise},$$

where T is the triangle bounded by lines x = 0, y = 0, and x + y = a (with a > 0).

(i) Find the mean life times EX and EY.

(ii) Consider now the simple queue with exponential service and Poisson input. Assuming that these different distributions have the same means as those in i above, which of EX and EY should correspond to the mean inter-arrival time and to the mean service time, when it is required that the system be in equilibrium? Show that the mean waiting time is then a/4.

40. Let X be a life time with exponential distribution with mean $1/\lambda$. For a fixed integer n > 1, define the life time M by

$$M = \min(X_1, \ldots, X_n)$$

where $X_1, \ldots, X_n$ are i.i.d. life times with a common distribution to that of X.

Consider now that simple queue with inter-arrival times having the same distribution as X.

(i) Justufy that M may be taken as the service time.

(ii) Deduce that this queueing system will be always in equilibrium, with $\rho = 1/n$.

(iii) Suppose that the cost function associated with the number of customers Y_t is given by

$$\phi(j) = m(-1)^j, \qquad \text{for } j = 0, 1, 2, \ldots, m > 0.$$

Find the average cost, and show that it tends
to m when n → ∞.

(iv)   Suppose that the cost function associated with
the waiting time W is given by

$$\phi(t) = n^2 t, \qquad \text{for } t \geq 0.$$

Find the average cost, and show that it tends
to $\mathbb{E}X$ when n → ∞.

41.   Consider the queueing system obtained from the simple
queue by requiring a finite waiting room of size r.
The number of customers in the system is described by
the birth and death process with coefficients

$$\lambda_i = \lambda \qquad \text{for } i = 0, 1, \ldots, r, \ \lambda_{r+1} = 0.$$

$$\mu_i = \mu \qquad \text{for } i = 1, 2, \ldots, r + 1, \ \mu_o = 0.$$

(a)   Show that the equilibrium solution P(j) for
distribution of customers is given by

$$P(j) = P(0)\rho^j \qquad \text{for } j = 0, 1, \ldots, r + 1$$

where P(0) is a normalizing constant and $\rho = \lambda/\mu$. Assume $\rho \neq 1$.

(b)   Assume the strict order service, and show that
the waiting time of a customer who on his
arrival meets i customers in the system has the
(complementary) d.f. of the form

$$D_i^c(t) = \int_{\mu t}^{\infty} x^{i-1} e^{-x} dx/(i - 1)!, \qquad 1 \leq i \leq r.$$

Deduce that the waiting time has the (comple-

mentary) d.f. given by

$$W^C(t) = \sum_{i=1}^{r} P(i) \, D_i^C(t).$$

(c)   Show that $P(r + 1)$ is the pr. of loss.

5
Inference from Interference

In our discussions of stochastic models earlier in
this book we obtained explicit expressions for probabilities
of interest (in a more or less convenient form) and restricted our attention to the study of analytical properties of
these formulae. That indeed was our objective from the very
beginning. In the first two chapters we obtained distributions of life times, discrete and continuous, whereas in the
chapter on renewals we encountered mixed cases and in Chapter 4, time-dependent distributions were discussed. Thus,
if X is a life time (or more generally, an r.v) its distribution function F fully describes the behavior of X.

As we have noticed, the d.f. F usually involves some
parameters. For example, in exponential distribution one
has λ, in gaussian distribution there are two parameters μ
and σ^2, and there are also two in binomial, n and p. These
parameters are specified by the model under consideration

and in general are related to some characteristic of F (such
as mean or variance). For simplicity, denote these parameters by the symbol θ—which may be a vector—and to stress
this dependence write $F(x,\theta)$ for the probability of the
event $X \leq x$. Thus, for the exponential distribution $\theta = \lambda$,
whereas for the gaussian distribution $\theta = (\mu,\sigma^2)$.

In this chapter our outlook will be basically different from that in earlier chapters. Having completed a
probabilistic study of our model, we wish now to associate
this model with observations of the behavior of real life
situations (described by our model). In other words, we turn
now to Statistics, that is, to a mathematical discipline
that allows us to draw conclusions from a collection of data.
Statistical methods provide reasonable procedures for such
a task, taking into account data obtained from observation.
It is this link between a mathematical model and concrete
data that is the basis of statistical reasoning. In other
words, probabilistic aspects restrict attention to analysis
of a model only, whereas statistical inference concerns a
mutual relationship between a model and available data. In
this sense, numerical aspects provided by data interfere
with the analytical aspects of our model—hence the title
of this chapter.

The connection of a probability model with real data
—the essence of statistical analysis—may now be illustrated as follows. Given a real life situation and the observ-

ed data (collected from observations of the situation), the statistical problem will consist of

(i)   Determination of the unknown distribution F

(ii)   If F is known (or assumed), determination of unknown θ

This corresponds basically to the division of statistical methods into non-parametric and parametric methods, respectively. In the first case, one selects an appropriate form of F (say, exponential) from a class of possible F, and in the second case one looks for the numerical value of parameter θ. As already noted, the real data should profoundly "interfere" with whatever beautiful theoretical models one can propose.

In this chapter we shall restrict our attention to parametric methods, and our discussion will be concerned with the estimation of parameters. Even so, we shall restrict ourselves to only one method, a very popular one, namely that of maximum likelihood. We shall look at simple estimation problems concerning typical distributions we encountered earlier in this book. Finally, we shall venture into the rather new territory of parameter estimation for processes, especially birth and death processes. Regrettably, we will be unable to touch the important area of testing of statistical hypotheses as it requires rather greater effort on the part of the reader as well as the writer.

Section 27: Esteemed Estimators

We shall now try to use data collected from observations to estimate unknown parameters in a distribution whose
form F is known. Let, then, X be a life time describing a
real life situation, and suppose that we made n observations
of X. This means that we recorded n values, say, $x_1, x_2, \ldots,$
x_n that X assumed. For example we may observe the waiting
time for a bus on n consecutive days, or durations of n
interarrival times in a Poisson process, or the lifetimes of
n consecutive periods between renewals, and so on. The
numbers $x_1, \ldots, x_n$ obtained in this way constitute our data.
How can we use it to determine the value of an unknown
parameter θ in the distribution $F(x, \theta)$ of X?

We could of course select a "good-looking" number
from our sequence $x_1, \ldots, x_n$ and declare it to be the
required value of θ. Or we can perform some manipulations
of our data, such as taking the average $(x_1 + \cdots + x_n)/n$,
and take the resulting number for θ. Why not? Well, such
arbitrary procedures do not seem to be reasonable, and we
have no assurance that they would yield results that even
remotely resemble the true but unknown value of θ. What is
needed is a procedure that is reasonable and that would
guarantee good approximation to the true value. Moreover,
we wish that our procedure will be valid not only for the
data we obtained but also for any other data (concerning
the same X) that may be obtained at different times and by
different people.

For this purpose one must consider a collection of random variables $X_1, X_2, \ldots, X_n$, each of which represents the outcome of observation of life time X. In this way we account for all possible observations of X of length n. It is natural to assume that these $X_i, i = 1, \ldots, n$, are i.i.d. (independent and identically distributed) with common distribution $F(x, \theta)$. Such a collection $(X_1, \ldots, X_n)$ is called a <u>random sample of size n</u>, where n is a positive integer.

To estimate the parameter on the basis of a random sample, we may regard (temporarily at least) the parameter as a life time, which is a certain function of a random sample. Thus we need to introduce a cost function h and to look at a random variable $h(X_1, \ldots, X_n)$. This is precisely a combination of life times, a topic we discussed extensively earlier in this book. The random variable $h(X_1, \ldots, X_n)$ is called a <u>statistic</u> (sometimes this name is applied to the cost function h), and it represents the parameter, regarded as a random variable. It is customary to write $\hat{\theta}$ for parameter interpreted in this manner, so

$$\hat{\theta} = h(X_1, \ldots, X_n).$$

As the procedure is intended to estimate the unknown parameter θ, it is customary to call $\hat{\theta}$ the <u>estimator</u> of θ. It should be remembered, however, that $\hat{\theta}$ is a random variable whereas θ is a number; note also that function h cannot depend on θ.

When h is chosen, that is, when the estimator $\hat{\theta}$ is known, we then return to our data $x_1, \ldots, x_n$ and evaluate $h(x_1, \ldots, x_n)$. The number obtained in this way is called the estimate of θ and is taken as the value of θ obtained from data:

$$\theta_0 = h(x_1, \ldots, x_n).$$

Care should be taken to distinguish between the estimator $\hat{\theta}$ (which is a random variable) and its value θ_0 (which is a number).

How do we select an appropriate statistic? It seems that we relocated the difficulty by allowing arbitrariness in the choice of h. Fortunately, the situation is not as bad as it may look. We shall soon discuss reasonable and proper methods for the selection of a suitable statistic. First, however, let us list two desirable properties of estimators.

As the estimator $\hat{\theta}$ is a random variable, it has its own distribution (as we have seen already), so we may calculate its mean value $\mathbb{E}\hat{\theta}$. The estimator $\hat{\theta}$ is said to be unbiased when its average equals the parameter

$$\mathbb{E}\,\hat{\theta} = \theta.$$

This condition is rather natural and is imposed frequently.

We would like also our estimators to be better when more data are provided. Thus when the sample size n increases, the estimator should be closer to the true value,

$$\hat{\theta} \to \theta \qquad \text{as } n \to \infty .$$

This is precisely the type of convergence we treated in Section 11. The estimator with the above property is said to be <u>consistent</u>.

<u>Example 1</u>: Let X be a life time with mean $\mathbb{E}(X) = \mu$. Suppose that we wish to estimate μ, so $\theta = \mu$. It is rather natural to select as our estimator $\hat{\theta}$ the sample mean

$$\overline{X}_n = (X_1 + \cdots + X_n)/n .$$

We know that $\mathbb{E}(\overline{X}_n) = \mu$ and that $\overline{X}_n \to \mu$ as $n \to \infty$ (as noted in Section 10). Hence, $\overline{X}_n$ is unbiased and a consistent estimator of μ. If now the values of n observations of the life time X are $x_1, \ldots, x_n$, then our estimate of μ is now

$$\mu_0 = (x_1 + \cdots + x_n)/n .$$

If required, we may find distribution of the sample mean $\overline{X}_n$, and it may be of interest to compute

$$pr(|\overline{X}_n - \mu_0| > \varepsilon) .$$

<u>Example 2</u>: In the same fashion, if the variance of X is the unknown parameter, $\theta = \sigma^2$, it is natural to take as an estimator the <u>sample variance</u>, defined as

$$s^2 = (1/n) \sum_{i=1}^{n} (X_i - \overline{X}_n)^2 .$$

It is not difficult to verify that $\mathbb{E}(s^2) = \sigma^2(n - 1)/n$, so s^2 is biased. Hence, replacing s^2 by $s^2n/(n-1)$, one

obtains the unbiased estimator. It is also known that both

these estimators are consistent.

If data $x_1,\ldots,x_n$ are available, direct (and perhaps

tedious) computation yields the estimate σ_0^2 of σ^2.

Remark: Both examples illustrate the method of moments in

which population moments (like mean and variance) are equat-

ed with sample moments (like $\overline{X}$ and s^2). Although we have

treated only the case when θ is a single parameter, the

analysis can be extended to several parameters. The follow-

ing example illustrates the procedure.

Example 3: Let X be the number of successes in Bernoulli

trials whose distribution has two parameters N and p. We

wish to find estimators $\hat{N}$ and $\hat{p}$ of these parameters, so now

$\theta = (N,p)$. We know that mean and variance of X are, respec-

tively Np and Np(1 - p), so equating population moments

with sample moments, we get the equations

$$\hat{N}\hat{p} = \overline{X}_n \qquad \text{and} \qquad \hat{N}\hat{p}(1 - \hat{p}) = s^2,$$

which can be solved for $\hat{N}$ and $\hat{p}$ in terms of $\overline{X}_n$ and s^2

(whose values are then computed from data $x_1,\ldots,x_n$).

(Warning: Do not confuse the number of trials N with the

sample size n.)

Note: The method described here produces a single number

for a value of the unknown parameter. Such a procedure is

called a point estimation, in contrast to interval estimation, which produces an interval such that we may say, with
confidence, that is contains the unknown parameter.

Section 28: As You Like It

Let us return now to the discussion of Bernoulli
trials (Section 11) and suppose that we wish to estimate
the probability p of a success. After all, we used that p
all the time previously, so it would be nice to know how we
can find its value. Well, the method of moments will do,
but we now plan to be slightly more sophisticated.

Again we perform n observations of a random variable
X that assumes values of 0 or 1, with 0 standing for failure
F and 1 for success S. This gives us a sequence of numbers
$x_1, \ldots, x_n$, where each x_i is either 0 or 1. Clearly, $x_1 +$
$\cdots + x_n$ is the total number of 1's (i.e., number of S's)
in n observations and $n - (x_1 + \cdots + x_n)$ is the number of
failures F (i.e., number of 0's).

We now ask the important question: how likely is to
obtain that set of numbers $x_1, \ldots, x_n$? That is, what is the
probability that consecutive n observations indeed yield
$x_1, \ldots, x_n$? The answer is immediate. As before, we represent the results of observations by independent identically
distributed r.v.'s, each taking values 0 and 1 with pr. 1 -
p and p, respectively. Hence the required pr. is

$$pr(X_1 = x_1, \cdots, X_n = x_n) = p^{x_1}(1 - p)^{1-x_1} \cdots p^{x_n}(1 - p)^{1-x_n}.$$

So far so good, but p is still unknown. Now comes the punch line. We look at the above pr. as a function of p, where $0 \leq p \leq 1$. If it is small, then the data $x_1, \ldots, x_n$ may be thought a rather unlikely occurrence; if the value is large, then numbers $x_1, \ldots, x_n$ may be regarded as what you would expect to obtain. Hence the logical step is to find such p that would maximize the above pr. of occurrence and take this p as our estimate.

Thus, denoting the above pr. by L(p), so in our case,

$$L(p) = p^{x_1 + \cdots + x_n}(1 - p)^{n-(x_1 + \cdots + x_n)},$$

the problem is to maximize L(p). To stress the point that we treat the above expression not as a probability but as a function of p, we call L(p) a likelihood function. Remember, L(p) is not a pr. in p, it is pr. in $x_1, \ldots, x_n$— hence the change in name.

Apart from two border cases when $x_1 + \cdots + x_n$ is 0 or n, we have L(0) = 0 and L(1) = 0. Since $L(p) \geq 0$, so there must be at least one maximum. If fact there is only one, as is seen from differentiation. Evidently, derivatives of L(p) and of log L(p) vanish at the same point, so in this case it is more convenient to "take log" first and then differentiate. Thus,

$$\log L(p) = (x_1 + \cdots + x_n) \log p$$
$$+ [n - (x_1 + \cdots + x_n)] \log (1 - p),$$

and

$$d \log L(p)/dp = (x_1 + \cdots + x_n)/p$$
$$- [n - (x_1 + \cdots + x_n)]/(1 - p).$$

Equating this to 0, the value of p that maximizes L(p) is found to be

$$p_m = (x_1 + \cdots + x_n)/n.$$

This is now taken as an estimate of p, and the corresponding estimator, to be called the maximum likelihood estimator (MLE), is again the sample mean:

$$\hat{p} = (X_1 + \cdots + X_n)/n.$$

Indeed when tossing a coin 100 times we get 52 heads and 48 tails; the MLE estimate of p is 0.52.

As a second example, suppose that the life time X has exponential distribution with unknown parameter λ, which we wish to estimate by the maximum likelihood method. As in the continuous case, pr. of a r.v. assuming a fixed value is zero, it is natural to pass to densities. So if the sample values are $x_1, \ldots, x_n$, then the likelihood function is now

$$L(\lambda) = \lambda^n e^{-\lambda(x_1 + \cdots + x_n)}$$

so

$$\log L(\lambda) = n \log \lambda - \lambda(x_1 + \cdots + x_n)$$

and

$$d \log L(\lambda)/d\lambda = n/\lambda - (x_1 + \cdots + x_n).$$

Hence the maximum occurs at

$$\lambda_m = n/(x_1 + \cdots + x_n).$$

Consequently the MLE of λ is

$$\hat{\lambda} = 1/\overline{X},$$

the reciprocal of the sample mean $\overline{X}$. This agrees with the method of moments, because $\mathbb{E}X = 1/\lambda$. It may be useful to point out that in the present case, λ ranges from 0 to infinity and $L(0) = 0$ and $L(\infty) = 0$. As $L(\lambda)$ is positive, there is no need to check the second derivative in order to verify that λ_m indeed gives maximum. If, for example, four observations of a certain life time measured in months give values 13, 20, 14, and 17, the sample mean is 16 so the estimate of λ is 1/16.

We can now summarize the maximum likelihood method of estimation. If θ is an unknown parameter in the distribution of a r.v. X, the first step is to write the likelihood function $L(\theta)$ based on the data $x_1, \ldots, x_n$. Thus by definition,

$$L(\theta) = f(x_1, \theta) \cdots f(x_n, \theta) = \prod_{i=1}^{n} f(x_i, \theta)$$

in the continuous case, and in the discrete case,

$$L(\theta) = p(x_1, \theta) \cdots p(x_n, \theta) = \prod_{i=1}^{n} p(x_i, \theta)$$

The next step is to find such θ that maximize $L(\theta)$. This θ_m is therefore a function of values $x_1, \ldots, x_n$, say $\theta_m = h(x_1, \ldots, x_n)$. The MLE is then taken to be

$$\hat{\theta} = h(X_1, \ldots, X_n).$$

In the above, the parameter θ may be a vector, say $\theta = (\theta_1, \theta_2)$.

Let us now work out a few interesting examples (see also Problems in this chapter).

****

Gaussian example: Suppose that the random variable X has gaussian distribution with mean μ and variance σ^2 and that both parameters μ and σ^2 are unknown. We wish to find their MLE based on a random sample of size n. Thus, $\theta = (\mu, \sigma^2)$, and using data $x_1, \ldots, x_n$, the likelihood function $L = L(\mu, \sigma^2)$ is easily found to be

$$L = \prod_{i=1}^{n} f(x_i) = (2\pi\sigma^2)^{-n/2} e^{-A(\mu, \sigma^2)}$$

where

$$A(\mu, \sigma^2) = (1/2\sigma^2) \sum_{i=1}^{n} (x_i - \mu)^2$$

and $f(x)$ is the gaussian density (from Section 9). Hence,

$$\log L = -A(\mu, \sigma^2) - (n/2) \log 2\pi\sigma^2.$$

As there are two parameters, we must take partial derivatives to find a maximum, so

$$\partial \log L / \partial \mu = \sum_{i=n}^{n} (x_i - \mu) / \sigma^2,$$

$$\partial \log L / \partial \sigma^2 = A(\mu, \sigma^2) / \sigma^2 - n(2\sigma^2)^{-1}$$

Equating these partial derivatives to zero, one finds

$$\mu = (x_1 + \cdots + x_n)/n, \qquad \sigma^2 = \sum_{i=1}^{n} (x_i - \mu)^2/n.$$

It is therefore easily seen that the MLE of mean μ is the sample mean $\overline{X}$. However, to obtain MLE of variance σ^2, it should be remembered that the estimator cannot involve the unknown parameter, so μ must be replaced by its estimate. Thus the MLE of σ^2 is the sample variance S^2 (we know from Section 26 that S^2 is a biased estimator). Thus our solution is

$$\hat{\mu} = (X_1 + \cdots + X_n)/n, \qquad \hat{\sigma}^2 = \sum_{i=1}^{n} (X_i - \overline{X})^2/n$$

It is of interest to check how the answer changes when one of the parameters is known. In the special case when variance σ^2 is known, the MLE of μ is obviously the same, i.e., the sample mean $\overline{X}$. On the other hand, if mean μ is known, the MLE of σ^2 is the unbiased estimator

$$\sum_{i=1}^{n} (X_i - \mu)^2/n.$$

<u>Uniform life time</u>: The popular uniform distribution exhibits a feature which may be puzzling at first sight. The density of the life time X is $f(x) = 1/\theta$ (for $0 < x < \theta$), and we wish to estimate θ. The likelihood function based on observations $x_1, \ldots, x_n$ is clearly

$$L(\theta) = (1/\theta)^n.$$

Now, differentiation makes no sense, and what is even worse, $L(\theta)$ does not include observed values $x_1, \ldots, x_n$. To solve this puzzle, we should look back at maximum likelihood procedure. The rule says "maximize," not "differentiate." We must find maximum by other means.

As the life time X takes values only between 0 and θ, the observed values should also fall in this range. It is now obvious that $\theta_0 = \max(x_1, \ldots, x_n)$ will maximize $L(\theta)$. Indeed, for $\theta > \theta_0$, evidently $L(\theta) < L(\theta_0)$. Thus, the MLE of θ is

$$\hat{\theta} = \max(X_1, \ldots, X_n).$$

We already encountered this life time in Section 6, where we found that

$$\mathbb{E}\,\hat{\theta} = n(n + 1)^{-1}\theta,$$

so $\hat{\theta}$ is a biased estimator of θ. We may again replace it by an unbiased estimator $\theta^* = \hat{\theta}(n + 1)/n$.

In contrast, the method of moments yields the unbiased estimator $\tilde{\theta} = 2\overline{X}$, which differs from $\hat{\theta}$. For comparison purposes, we can easily calculate the variance of both unbiased estimators:

$$\text{var } \theta^* = \theta^2/n(n+2), \qquad \text{var } \tilde{\theta} = \theta^2/3n.$$

Clearly, $\text{var } \theta^* < \text{var } \tilde{\theta}$, so in this sense the MLE θ^* is better than $\tilde{\theta}$.

Section 29: Estimating Poisson Input

In Chapter 4 we considered stochastic processes—essentially of the birth and death type—and observed that their distributions involve some parameters and also depend on time t. Although estimation of these parameters may be carried out according to the same principles described earlier, the task becomes more complicated.

In this section we shall examine the Poisson input $(Y_t, 0 \le t < \infty)$, described at length in Section 16 (as a renewal process) and in Section 19 (as a birth process). Here Y_t denotes a state of the system at time t, and $N_t = Y_t - Y_0$ represents the number of arrivals up to time t. We shall take $Y_0 = 0$ (i.e., there are no customers initially) so $N_t = Y_t$; when convenient, we shall write N(t) in place of N_t. Recall that we can look at the Poisson input process by observing the number of arrivals in a fixed period or the length of time needed for a fixed number of arrivals. This duality, already discussed in Sections 16 and 19, is responsible for different approaches to estimation problems.

Suppose that we wish to estimate the intensity λ of the Poisson input. Recall that λ is a rate of arrivals (measured in customers per unit of time) as well as the reciprocal of the mean arrival time. Here of course, the interarrival times are i.i.d. and have exponential distribution. Our approach will be through the use of the maximum likelihood method.

29.1.

Suppose that we made n observations of the Poisson input at instants $0 < t_1 < \cdots < t_n$ and recorded the values $0 \leq j_1 \leq \cdots \leq j_n$, respectively. Let us calculate the probability of obtaining the sequence $j_1, \ldots, j_n$ at instants $t_1, \ldots, t_n$. Using properties of the Poisson process and the form of its transition probabilities (see Section 19), this joint distribution is

$$pr(N(t_1) = j_1, \ldots, N(t_n) = j_n) = pr(N(t_1) = j_1, N(t_2)$$

$$- N(t_1) = j_2 - j_1, \ldots, N(t_n) - N(t_{n-1}) = j_n$$

$$- j_{n-1})$$

$$= pr(N(t_1) = j_1) \, pr(N(t_2) - N(t_1) = j_2 - j_1)$$

$$\cdots \, pr(N(t_n) - N(t_{n-1}) = j_n - j_{n-1})$$

$$= \prod_{h=1}^{n} \frac{[\lambda(t_h - t_{h-1})]^{j_h - j_{h-1}}}{(i_h - i_{h-1})!} e^{-\lambda(t_h - t_{h-1})}$$

where $t_0 = 0$, $j_0 = 0$. As required by the method, we take the above probability as our likelihood function for λ. Obvious simplification yields

$$L(\lambda) = C\lambda^{j_n} e^{-\lambda t_n}$$

where C is a term that does not involve λ. The next step is maximization of $L(\lambda)$. Proceeding as before

$$\log L(\lambda) = \log C + j_n \log \lambda - \lambda t_n,$$

and differentiation yields the estimate $\lambda = j_n/t_n$, which depends only on the last observation.

As time t_n has been fixed, the MLE of λ is

$$\hat{\lambda} = N(t_n)/t_n.$$

Hence:

$$\mathbb{E} \hat{\lambda} = \lambda, \qquad \text{var } \hat{\lambda} = \lambda/t_n,$$

so $\hat{\lambda}$ is an unbiased estimator.

Thus, if the last observation occured at the instant $t_n = 2$ hours and recorded $N(t_n) = 16$ customers present, then the estimate of the rate λ is $16/2 = 8$ customers/hour.

29.2

In the above approach we really did not take into account that interarrival intervals are themselves random variables. In order to utilize complete information, it is necessary to look at the path (the sample function) of the process, namely to record both jumps (arrivals) and time distance between arrivals. This is achieved with help of the so-called <u>sample function density</u>, introduced below.

Consider a time interval $[0,t)$ of fixed length t, and suppose that there have been recorded n arrivals up to time t. Let $x_1,\ldots,x_n$ be the observed durations of the consecutive interarrival intervals. In the notation of renewal theory, X_i, $i = 1,\ldots,n$ are inter-arrival intervals, and the instants of arrival are

$$S_i = X_1 + \cdots + X_i, \qquad i = 1,\ldots,n \ (S_0 = 0)$$

With N_t denoting the number of arrivals up to time t, we obviously have

$$(S_n \leq t < S_{n+1}) = (N_t = n).$$

As life times X_i are independent and have the same exponential distribution, it is clear that the joint density of arrival instants $S_1,\ldots,S_n$ has the form

$$a_n(s_1,\ldots,s_n) = \prod_{i=1}^{n} \lambda e^{-\lambda x_i} = \lambda^n e^{-\lambda s_n}$$

for $0 < s_1 < \cdots < s_n$, where $s_i = x_1 + \cdots + x_i$ $(i = 1,\ldots, n)$ are values of S_i. The sample function density $g_t(n,s_1, \ldots,s_n)$ determines the probability of obtaining a particular path of the process on $[0,t)$ with $N_t = n$ points located for $n \geq 1$ at times $S_1 = s_1,\ldots,S_n = s_n$. This means that

$$g_t(n,s_1,\ldots,s_n) = \text{pr}(N_t = n|S_1 = s_1,\ldots,S_n)a_n(s_1,\ldots,s_n)$$

$$\text{for } n \geq 1,$$

and

$$g_t(0) = e^{-\lambda t} \qquad \text{when } n = 0.$$

However, by the Markov property,

$$pr(N_t = n | S_1 = s_1, \ldots, S_n = s_n) =$$

$$pr(S_n \leq t < S_{n+1} | S_n = s_n)$$

$$= pr(X_{n+1} > t - S_n | S_n = s_n) = e^{-\lambda(t-s_n)}.$$

Consequently,

$$g_t(n, s_1, \ldots, s_n) = \lambda^n e^{-s_n \lambda} e^{-\lambda(t-s_n)} = \lambda^n e^{-\lambda t}.$$

Observe that integration of the above density over the region $0 \leq s_1 \leq \cdots \leq s_n \leq t$ yields the Poisson distribution for $pr(N_t = n)$; see Problem 12.

Taking density g_t to be our likelihood function for λ:

$$L(\lambda) = \lambda^n e^{-\lambda t}$$

we have as before

$$\log L(\lambda) = n \log \lambda - \lambda t$$

so the maximum occurs at $\lambda = n/t$. Consequently, the MLE of λ is

$$\hat{\lambda} = N_t/t$$

with $\mathbb{E}\hat{\lambda} = \lambda$. Thus, for example, if the process has been observed for 10 hours to produce 15 arrivals, the estimate of the rate λ would be 1.5 customers per hour.

Observe that the estimator obtained here is essentially the same as that by the method of counts (this is

due to special properties of the Poisson process but need
not be true in general).

Section 30: Estimating the Birth and Death Process

In Section 25 we considered a birth and death process
$(Y_t, 0 \leq t < \infty)$ used to describe fluctuations in a population size occurring with time, such as a number of customers
in some system. Thus, Y_t denotes the size of population
(i.e., the state) prevailing at time t. As already explained, in the birth and death process changes of states during
a short interval can take place only to the immediate
neighboring states. Thus, if at a given instant of time
the system is in state i, then a transition can occur only
to state i + 1 (with birth rate λ_i) or to state i - 1 (with
death rate μ_i), or the system may remain in state i (with
rate $\lambda_i + \mu_i$). Recall that our state space is a set of non-
negative integers and that $\mu_0 = 0$. We shall assume now that
there are no absorbing states so $\lambda_i + \mu_i > 0$ for all $i \geq 0$.

Although we did not show it explicitly, the Markov
property implies that duration of a state i has exponential
distribution with mean $1/(\lambda_i + \mu_i)$; we hinted at this
possibility when discussing the birth process in Section 24
(see also Problem 14). It will be convenient to treat up
and down transitions simultaneously, so we shall write
q(i,j) for the rate of transition from state i to state j:

$$q(i,i + 1) = \lambda_i, \qquad q(i,i - 1) = \mu_i,$$

$$q(i) = \lambda_i + \mu_i.$$

Thus, duration of state i has exponential density with a parameter q(i). Furthermore, the probability of a jump from state i to state j, denoted by r(i,j), is defined as

$$r(i,j) = q(i,j)/q(i) \qquad \text{for } i \neq j \text{ and } r(i,i) = 0.$$

Hence,

$$r(i,i + 1) = \lambda_i/(\lambda_i + \mu_i) \qquad \text{for } i \to i + 1$$

$$r(i,i - 1) = \mu_i/(\lambda_i + \mu_i) \qquad \text{for } i \to i - 1 \text{ for } i \geq 1,$$

and

$$r(0,1) = \lambda_0/\lambda_0 = 1 \qquad \text{for } 0 \to 1 \text{ when } i = 0.$$

Suppose now that intensities λ_i and μ_i are functions of a parameter θ (which may be a vector with several components). For example, in the M/M/1 queue with $\lambda_i = \lambda$ and $\mu_i = \mu$, we can consider two-dimensional parameter $\theta = (\lambda,\mu)$. In statistical problems, the parameter θ is unknown and our task will be to estimate θ. We shall use again the maximum likelihood method. The procedure is essentially the same as that described in the previous section on Poisson input, but analytical complications inevitably arise. As before, we shall look at possible realizations of the birth and death process and determine the sample function density. This will be our likelihood function $L = L(\theta)$. The next step will be maximization of $L(\theta)$.

We shall first develop the general formula for the birth and death process and then simplify it a bit. Afterward we shall apply our results to the M/M/1 queue and to a birth process.

30.1.

Consider now a time interval $[0,t)$ of fixed length t, and suppose that continuous observation produced n transitions (jumps). Let $S_1 < S_2 < \cdots < S_n$ be the random times at which these transitions occurred; define also $S_0 = 0$ (the initial instant). Denote by X_k the time interval between the k-th and the (k + 1)-th jump, so

$$X_k = S_{k+1} - S_k > 0 \qquad k = 0, 1, \ldots$$

or equivalently

$$S_k = X_0 + \cdots + X_{k-1}, \qquad k = 1, 2, \ldots .$$

Let N_t be the number of transitions (jumps) up to time t. One has evidently

$$(N_t = n) = (S_n \leq t < S_{n+1}), \qquad n = 0, 1, 2, \ldots .$$

Although this situation resembles that of renewals, it should be stressed that in general the life times X_k are not independent (and have exponential distribution, but with different parameters).

Next, consider a family of random variables $(Y_k, 0 \leq k < \infty)$ defined by

$$Y_k = Y(S_k), \qquad k = 0, 1, 2, \ldots .$$

Thus, Y_k represents the state of the system immediately after the k-th jump. It is clear that the behavior of the whole system is described by pairs (Y_k, X_k), i.e., by the magnitude at jumps and the distance between jumps. Observed values of (Y_k, X_k) will constitute our data. As customary, we shall use lowercase letters (with subscripts, if necessary) to denote the observed values of random variables.

Suppose that the system is initially in state i_0 and stays there till instant s_1 when the (first) jump occurs to state i_1 (which is either $i_0 + 1$ or $i_0 - 1$). State i_1 lasts till instant s_2 of the (second) jump to state i_2. The procedure continues till the n-th jump to state i_n, occurring at time s_n before t. See Figure 30.1. This accounts for n transitions up to time t. We thus observed a path or a sample function of a birth and death process. How do we determine the probability of such a path? Note that Y_k are discrete random variables, whereas S_k are continuous life

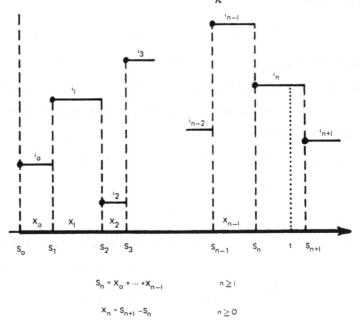

$$S_n = X_0 + \cdots + X_{n-1} \qquad n \geq 1$$

$$X_n = S_{n+1} - S_n \qquad n \geq 0$$

Figure 30.1

times so events $(S_k = s_k)$ have probability zero. This
explains the rather cumbersome notation for the density
(for a similar situation, see the discussion of density in
Section 2).

We shall use again the sample function density
appropriate for such mixed situations. We shall rely on
the Markov property in a rather intuitive fashion—thus in
conditioning a transition at time S_n does not depend on
preceding transitions. We shall also make use of equiva-
lence between sequences (S_k) and (X_k).

30.2.

In order to derive the expression for the sample
function density, we shall proceed slowly but surely, using
a step-by-step argument that enjoys intuitive interpretation.

Suppose that a system is in initial state i_0 and that
at instant $s_1 = x_0$ a jump to state i_1 occurs with probability $r(i_0,i_1)$, where i_1 is either $i_0 - 1$ or $i_0 + 1$. The
duration of state i_1 is exponential with mean $1/q(i_1)$, and
so is the duration of state i_0 with mean $1/q(i_0)$. Assuming
that only this one transition $i_0 \rightarrow i_1$ takes place before
time t, we can compute the following probability:

$$\text{pr}(Y_0 = i_0,\ Y_1 = i_1,\ N_t = 1,\ X_0 \leq x_0) =$$

$$\text{pr}(Y_0 = i_0,\ X_0 \leq x_0,\ X_0 \leq t < X_0 + X_1,\ Y_1 = i_1)$$

$$\int_0^{x_0} \text{pr}(Y_0 = i_0,\ X_0 \approx du)\text{pr}(Y_1 = i_1,\ t-u > X_1 | Y_0,\ X_0 = u)$$

$$= \int_0^{x_0} \Pi(i_0)q(i_0)e^{-q(i_0)u}\ du\ r(i_0,i_1)e^{-q(i_1)(t-u)}.$$

Differentiating this expression with respect to x_0, we obtain the expression for the sample function density for n = 1, which we take for our likelihood function:

$$L_1 = \Pi(i_0)e^{-q(i_0)x_0}q(i_0,i_1)e^{-q(i_1)(t-x_0)}.$$

Here $\Pi(i_0)$ is the initial distribution of Y_0.

Next consider the case of n = 2. With the first transition occurring at s_1 as before, the second transition will occur at instant $s_2 = x_0 + x_1$ to state i_2, which is either $i_1 - 1$ or $i_1 + 1$. Assuming that only these two transitions take place before time t, we can compute the following probability:

$$\text{pr}(Y_0 = i_0, \ Y_1 = i_1, \ Y_2 = i_2, \ N_t = 2, \ S_1 \leq x_0, \ S_2 \leq x_0 + x_1)$$

$$= \text{pr}(Y_0 = i_0, \ X_0 \leq x_0, \ Y_1 = -1, \ X_1 \leq x_0 + x_1 - X_0, \ Y_2 = i_2,$$

$$X_2 > t - X_0 - X_1)$$

$$= \Pi(i_0)q(i_0) \int_0^{x_0} e^{-q(i_0)u_0}\,du_0 + r(i_0,i_1) \int_0^{x_1} q(i_1)e^{-q(i_1)u_1}$$

$$du_1\, r(i_1,i_2)e^{-q(i_2)(t-u_0-u_1)}.$$

Differentiating with respect to x_0 and x_1, we obtain the expression for the sample function density for n = 2, which we take for our likelihood function:

$$L_2 = \Pi(i_0)e^{-q(i_0)x_0}q(i_0,i_1)e^{-q(i_1)x_1}q(i_1,i_2)e^{-q(i_2)(t-x_0-x_1)}.$$

Proceeding in the same fashion, we calculate for an integer n the probability

$$W = pr(Y_0 = i_0, Y_1 = i_1, \ldots, Y_n = i_n, N_t = n,$$

$$S_1 \leq s_1, \ldots, S_n \leq s_n),$$

where $s_i = x_0 + \cdots + x_{i-1}$, and then compute the sample function density as $\delta^n W / \delta x_0 \cdots \delta x_{n-1}$, which we take for our likelihood function. We cheerfully skip the explicit calculations and content ourselves with quoting the final result:

$$L_n = \Pi(i_0) \left(\prod_{k=0}^{n-1} e^{-q(i_k) x_k} q(i_k, i_{k+1}) \right) e^{-q(i_n)(t-s_n)}$$

for $i_k \neq i_{k+1}$, $t \geq s_n$, and $n \geq 1$. On the other hand, for $n = 0$ one can take

$$L_0 = \Pi(i_0) e^{-q(i_0)t}.$$

It may be remarked that the above formula for L_n is valid for more general Markov chains than the birth and death process, with proper interpretation of intensities $q(i,j)$ and $q(i)$. However, we consider here only birth and death processes, as in Chapter 4. In particular, for the Poisson process $q(i, i + 1) = q(i) = \lambda$ for all i, and one obtains L, as already stated in Section 29.2. It is of interest to note that letting $t = s_n$ in the expression for L_n, one obtains density corresponding to time of continuous observation until n jumps (transitions) take place.

30.3.

The fact that in a birth and death process only transitions of the form

$$i \rightarrow i \pm 1$$

occur simplifies the above expression for the likelihood function L_n.

Consider again the case $n = 2$. The expression for L_2 stated above may be re-written as

$$L = \Pi(i)q(i)e^{-q(i)s_1}r(i,j)q(j)e^{-q(j)(s_2-s_1)}$$
$$\times\ r(j,k)e^{-q(k)(t-s_2)}$$

which reduces to

$$\Pi(i)e^{-q(i)s_1}\lambda_i e^{-q(i+1)(s_2-s_1)}\lambda_{i+1}e^{-q(i+2)(t-s_2)}$$

$$\text{for } i \rightarrow i + 1 \rightarrow i + 2$$

or

$$\Pi(i)e^{-q(i)s_1}\lambda_i e^{-q(i+1)(s_2-s_1)}\mu_{i+1}e^{-q(i)(t-s_2)}$$

$$\text{for } i \rightarrow i + 1 \rightarrow i$$

or

$$\Pi(i)e^{-q(i)s_1}\mu_i e^{-q(i-1)(s_2-s_1)}\lambda_{i-1}e^{-q(i)(t-s_2)}$$

$$\text{for } i \rightarrow i - 1 \rightarrow i$$

or

$$\Pi(i)e^{-q(i)s_1}\mu_i e^{-q(i-1)(s_2-s_1)}\mu_{i-1}e^{-q(i-2)(t-s_2)}$$

$$\text{for } i \to i - 1 \to i - 2.$$

This provides the pattern for the general situation. Let a_j be the number of transitions $j \to j + 1$ (arrivals to state j), d_j be the number of transitions $j \to j - 1$ (departures from state j), and z_j be the total time spent in state j, with

$$\sum_j (a_j + d_j) = n, \qquad \sum_j z_j = t.$$

Then the sample function density of a sequence of n transitions up to time t may be written in the general form

$$L = \Pi(i) \prod_{j=0}^{\infty} \lambda_j^{a_j} \prod_{j=1}^{\infty} \mu_j^{d_j} \prod_{j=0}^{\infty} e^{-(\lambda_j+\mu_j)z_j}.$$

Observe that these products (over all states) contain an acutally finite number of terms, because a_j, d_j, and z_j are zero for states j not appearing in a sequence.

With the likelihood function L given above, we can now proceed as before. "Taking log" again, one has

$$\log L = \log \Pi(i) + \sum_{j=0}^{\infty} a_j \log \lambda_j + \sum_{j=1}^{\infty} d_j \log \mu_j$$

$$- \sum_{j=0}^{\infty} (\lambda_j + \mu_j)z_j.$$

Ignoring the initial state i, differentiation yields for a
vector $\theta = (\theta_k)$:

$$\delta \log L / \delta\theta_k = \sum_{j=0}^{\infty} a_j(\lambda_j'/\lambda_j) + \sum_{j=1}^{\infty} d_j(\mu_j'/\mu_j)$$

$$- \sum_{j=0}^{\infty} (\lambda_j' + \mu_j')z_j,$$

where the prime denotes differentiation with respect to
component θ_k. The estimates of θ_k are then obtained by
solving the system of equations

$$\delta \log L / \delta\theta_k = 0.$$

Unfortunately, the calculations may be rather involved,
except in some special cases. Two interesting examples are
given subsequently.

Remark: Suppose that we take all the intensities λ_j and
μ_j as parameters, so $\theta = (\lambda_j, \mu_j)$. Then clearly the estimates are given by

$$\lambda_j = a_j/z_j, \qquad \mu_j = d_j/z_j,$$

in agreement with intuition.

****

Note: As we have seen in Section 27, estimators are random
variables. Hence, let us rephrase our results in terms of
random variables.

Let $A_j(t)$ be a random variable representing the
number of transitions $j \to j + 1$ (arrivals) up to time t

(its observed values being a_j), and let $D_j(t)$ be a random variable representing the number of transitions $j \to j - 1$, (departures) up to time t (its observed values being d_j). Thus,

$$A(t) = A_j(t) \quad \text{and} \quad D(t) = D_j(t)$$

represent, respectively, the total number of arrivals and departures up to time t; write for their observed values

$$A(t) = \sum_j a_j \quad \text{and} \quad d = \sum_j d_j.$$

Evidently,

$$A(t) + D(t) = N_t, \quad a + d = n.$$

On the other hand, $S(N_t) \le t$ is the total time needed for N_t transitions to occur up to time t. Clearly, $t - S(N_t)$ is the remaining time since the last transition.

Let $Z_j(t)$ be the total time spent in state j up to time t, with observed values z_j, so

$$Z(t) = \sum_j Z_j(t) = t$$

This includes the interval of length $t - S(N_t)$ spent in the state after the last transition.

If instead of a fixed time t one continuously observes the system until exactly n transitions have occurred, we should replace t by a random time S_n (for fixed n).

30.4. Simple queue

As noted in Section 22, we have a birth and death

process with coefficients

$$\lambda_i = \lambda \qquad \text{for } i \geq 0, \qquad \mu_i = \mu \qquad \text{for } i \geq 1.$$

We now take

$$\theta_1 = \lambda \qquad \text{and} \qquad \theta_2 = \mu$$

and wish to estimate both parameters λ and μ.

The general formula yields for the likelihood L (neglecting the initial state)

$$\log L = \log \lambda \sum_{j=0}^{\infty} a_j + \log \mu \sum_{j=1}^{\infty} d_j - \lambda \sum_{j=0}^{\infty} z_j - \mu \sum_{j=1}^{\infty} z_j$$

$$= a \log \lambda + d \log \mu - \lambda t$$

$$- \mu(t - x_s).$$

Thus,

$$\delta \log L / \delta \lambda = a/\lambda - t = 0$$

$$\delta \log L / \delta \mu = d/\mu - (t - x_0) = 0$$

and the estimates are

$$\lambda = a/t, \qquad \mu = d/(t - x_0),$$

which correspond to estimators

$$\hat{\lambda} = A(t)/t \qquad \text{and} \qquad \hat{\mu} = D(t)/(t - X_0).$$

As noted earlier, if the observations are carried out till exactly n transitions occur, the time t should be replaced by the total observation time S_n. For example, if the continuous observation during 5 hours of a simple queue that is initially empty, but the first arrival occurs immediately after the observation starts, produced 70 arrivals and 30 departures, then estimates of rates are

$$\lambda = 70/5 = 14 \text{ customers per hour},$$

$$\mu = 30/5 = 6 \text{ customers per hour}.$$

30.5. Birth process

As noted in Section 24, we have only up transitions $j \rightarrow j + 1$, with birth coefficient λ_j defined for all $j \geq 0$ (and all μ_j vanish).

Suppose that there are initially i customers in the system, so $\Pi(i) = 1$. Subsequently n new customers arrive up to time t, so the system is in state $i + n$ at instant t. Evidently, now $a_j = 1$ for each j, and the duration of each state coincides with the inter-arrival time. Hence, in our notation (with malicious subscripts),

$$z_j = x_{j-i} \qquad \text{for } j = i, \ldots, i + n$$

and

$$t = \sum_{j=i}^{i+n} z_j = \sum_{k=0}^{n} x_k = s_n + z_{i+n}.$$

See Figure 30.2. Observe, however, that the interval between the last arival and the instant t, whose length has been denoted by z_{i+n}, also has exponential density (as noted in Section 3). Therefore, the likelihood function is now

$$L = \prod_{j=i}^{i+n-1} \lambda_j \prod_{j=i}^{i+n-1} e^{-\lambda_j z_j}, \qquad e^{-\lambda_{n_{ti}}(t-s_n)}$$

and

$$\log L = \sum_{j=i}^{i+n-1} \log \lambda_j - \sum_{j=1}^{i+n-1} \lambda_j z_j - \lambda_{i+n}(t - s_n).$$

Assuming now that λ_j depends on a single parameter θ, we find in the usual manner that

$$d \log L/d\theta = \sum_{j=i}^{i+n-1} \lambda_j'/\lambda_j - \sum_{j=i}^{i+n-1} \lambda_j' z_j - \lambda_{i+n}'(t - s_n)$$

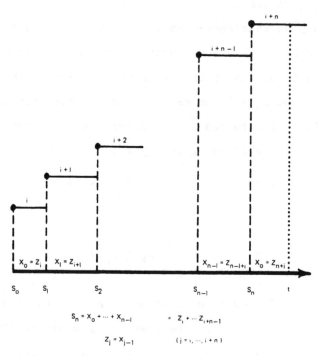

$$s_n = x_0 + \cdots + x_{n-1} \qquad = z_i + \cdots z_{i+n-1}$$

$$z_j = x_{j-1} \qquad (j = i, \ldots, i+n)$$

Figure 30.2

where $\lambda_j' = d\lambda_j/d\theta$. The estimate of θ is then obtained by solving $d \log L/d\theta = 0$. The following special cases are instructive.

(a) Linear input. Here $\lambda_j = j\lambda$ for $j \geq 1$, with $i = 1$. Take $\theta = \lambda$. Hence

$$d \log L/d\lambda = n/\lambda - \sum_{j=1}^{n} jz_j - (1 + n)(t - s_n) = 0,$$

so the estimate is

$$\lambda = n/w \qquad \text{where}$$

$$w = \sum_{j=1}^{n} jz_j + (1 + n)(t - s_n).$$

Recall that $z_j = x_j$ is the duration of state j and $t - s_n$ is the duration of state $n + 1$ (incomplete, and occurring after the n-th arrival).

Note that the quantity w is equal to the area under the graph of Y_s between 0 and t. So we can write the estimator of λ in the form

$$\lambda = N_t / \int_0^t Y_s \, ds.$$

(b) Poisson input. Here $\lambda_j = \lambda$ for all $j \geq 0$, with $i = 0$. Taking $\theta = \lambda$, we have

$$L = \lambda^n e^{-\lambda s_n} e^{-\lambda(t-s_n)} = \lambda^n e^{-\lambda t},$$

In agreement with that found in Section 29.

****

We now close our discussion of Statistics, before it takes us too far. We must remark, however, that we have touched only the tip of the iceberg.

Section 31: Modeling

Everything has its end (except a pencil, which has two ends), so the time has come to stop our discussion. Well, it is natural to take an extra moment to reflect on what we have done.

Our aim has been to show that highly complex "real life" problems involving random fluctuations, may be analyzed with sufficient accuracy by methods of theory of stochastic processes, which in part at least meet the needs of such problems. If at the same time we have painted a rather wearisome picture, no blame should be put on our subject; for after all, we have enjoyed fruits of scientific thought.

31.1

As already remarked in the Preface, we have constructed probabilistic models of real life problems under consideration and then have subjected our models to mathematical analysis and have compared our conclusions with actual observations. In this book, our models have been rather simple, because of the tools at our disposal, yet they have been sufficiently accurate to draw reasonable conclusions.

We have also noticed a rather important fact that often we considered not a single model but rather a class of models, applicable to diverse situations. This has been clearly indicated with the birth and death models, where diversity was expressed by the proper choice of coefficients (intensities). Similarily, renewal models have wide applicability. We have been also impressed by the generality of models based on first entrance times (absorption, waiting, busy periods, and so on).

What makes a model useful and reliable? This question has been addressed for a long time by many investigators. Of special interest in this respect is the recent book The Craft of Probabilistic Modeling, edited by J. Gani, which consists of a collection of the personal accounts of several researchers. The readers will find them very informative and penetrating, as well as entertaining. The present author may be forgiven by quoting from his article in Gani's book:

Mathematical modeling, like painting or photography, is an art, requiring proper balance between composition and the ability to convey a message. A good

mathematical model, aiming to present an idealistic
image of a real-life situation, should be accurate
as well as selective in its description, and should
use mathematical tools worthy of the problem.

The models we described here attempted to live up to the
above criteria, but our efforts have been restricted by our
objective to present a "first look." We sincerely hope,
however, that the effort by readers of this book has not
been in vain. We strongly recommend that interested readers
will continue their study of the subject by following the
Suggestions for Further Reading (Appendix C).

31.2.

Throughout this book we referred to stochastic
processes, but actually we did not specify what a stochastic
process is. At this stage, if you have read this far, you
should have a good intuitive feeling of what this is all
about.

It may be useful to add a few general comments on
stochastic (or random) processes without entering into
mathematical details. In the most general terms, a stochastic process is any family of random variables, say, $X =$
$(Y_t, t \in \Pi)$, where Π is an index set, usually a real line or
its subset, or a countable set. For example, the renewal
processes $(S_n, n = 0, 1, 2,...)$ and $(N_t, 0 \le t < \infty)$, discussed in Chapter 3, and the Markov process $(Y_t, 0 \le t < \infty)$

Chapters 4 and 5. In physical applications, the index set
usually represents time and the empirical approach describes
the stochastic process as any process in nature whose evolu-
tion in time depends on chance. The principal objects of
interest are properties of random variables X_t, representing
probabilistic conditions imposed on random variables. This
is often conveniently expressed by the joint distribution
functions for any finite collection of random variables:

$$\text{pr}(X_{t_1} \leq x_1, \ldots, X_{t_n} \leq x_n) = F_{t_1, \ldots t_n}(x_1, \ldots, x_n)$$

for any finite collection of indices $t_1, \ldots, t_n$ and for any
real numbers $x_1, \ldots, x_n$. Although this approach is important
for theoretical considerations, in its application considerable simplifications are imposed (which nevertheless encom-
pass wide classes of processes). On the one hand, we have
Markov processes, stationary processes, and martingales,
and on the other hand we have very special processes like
random walks, the Poisson process, Brownian motion, and
many others.

A stationary process is defined by the requirement
that its finite dimensional distributions are invariant
under shift, that is, they depend only on time differences,
not on the absolute time. In particular, one-dimensional
d.f's pr.$(X_t \leq x)$ cannot depend on t. We have seen several
instances of this "equilibrium condition."

As we explained in Chapter 4, Markov processes are
characterized by a lack of memory, in the sense that the

future development of the process depends only on the
present, not on the past. This property is expressed in
terms of conditioning as

$$\mathrm{pr}(X_t \leq x | X_u, \ u \leq s) = \mathrm{pr}(X_t \leq x | X_s), \qquad \text{for } s \leq t.$$

We have seen in Chapters 4 and 5 special cases of the above
relation, which lead us to the "balance equation"

$$p_{ij}(t + h) = \sum_k p_{ik}(t)p_{kj}(h)$$

where

$$p_{ij}(t) = \mathrm{pr}(X_t = j | X_0 = i).$$

Markov processes have enormously developed theory, a paradise for mathematicians, but at the same time have many
useful practical applications, as we have seen in our
"Markovian dance."

We did not explicitly use martingales, as their
presentation requires more sophisticated ideas. Yet martingales found their way into numerous applications, such as
queueing theory, control, prediction and filtering,
statistics, and many others. For the sake of an example,
consider a process

$$M_t = Y_t - \lambda t, \qquad 0 \leq t < \infty,$$

where (Y_t) is a Poisson process with intensity λ. In
Section 19 we calculated the conditional expectation, which
we now re-write in the form

$$\mathbb{E}\,(Y_t|Y_s) = \lambda(t - s) + Y_s, \qquad s \le t.$$

From this, using the "black magic" of conditioning, we obtain the martingale property

$$\mathbb{E}\,(M_t|M_u, u \le s) = M_s, \qquad s \le t.$$

We see that a martingale preserves conditional means, in the sense that the average of the future cost M_t, conditional on past and present, equals the present cost M_s.

****

As we have repeatedly pointed out, in all our discussions we tried to bridge a gap between theory and practice. Indeed, on the one hand, spectacular progress in the theory of stochastic processes, especially Markov processes and martingales, lead to widespread applications of probabilistic methods. On the other hand, "thinking with models" has become a common practice in applied fields.

In particular, Markovian models are entering new areas of applications. Although Markovian models are only approximate (by their very structure), more elaborate models involving semi-Markovian or point processes are important extensions of the Markov structure. We once again urge the reader to consult the Suggestions for Further Reading (Appendix C).

31.3.

As the last comment, we must add a warning to over-enthusiastic believers in mathematical models about the

dangers of interpreting mathematics in terms of external
non-mathematical concepts. Bitter experience indicates that
such unhealthy mixtures frequently lead to abuses of the
basic principles we described in this book and are sources
of misleading conclusions.

On the one hand, the mild versions of such trans-
gressions may even be amusing by their obvious nonsensical
character, but on the other hand the more serious abuses
may lead (and in fact did lead) to "religious wars" among
camps of believers.

In the first category we have school children jokes
(if you put one hand into ice-cool water and the other into
boiling water, then on average you are comfortable), propa-
ganda speeches (if elected, I shall see that all earnings
will be above average), and even some statistical statements
(the average size of an American family is two and half
children). The following quotation from a letter by Babbage
to the poet Tennyson (concerning "The Vision of Sin")
illustrates well effects of mixing unrelated ingredients
(as reported in the Mathematical Gazette):

> In your otherwise beautiful poem, there is a verse
> which reads: "Every moment dies a man, Every moment
> one is born." It must be manifest that, were this
> true, the population of the world would be at a
> standstill. In truth the rate of birth is slightly
> in excess of that of death. I would suggest that in

the next edition of your poem you have it read:
"Every moment dies a man, Every moment 1 1/16 is
born." Strictly speaking this is not correct. The
actual figure is a decimal so long that I cannot get
it in the line, but I believe 1 1/16 will be sufficiently accurate for poetry.

Let us now turn to the second category of abuses,
namely those more profound. The history of Probability
Theory provides several illustrations of the imposition of
non-methematical ideas on a mathematical model, thus leading
to serious consequences. We shall briefly describe three
such famous instances of mixing mathematics with philosophy
(albeit done in good faith). As seen today, these examples
may be dismissed as nothing but curiosities, yet their
impact on a way of thinking has been tremendous.

(a) <u>St. Petersburg paradox</u> (Daniel Bernoulli, 1738):
Consider a game of waiting for the first success in Bernoulli
trials (named after James, not Daniel) with $p = 1/k$, where
k is a positive integer larger than 1 ($k = 2$ is the classical value). If the first sucess occurs at the n-th trial,
the player receives k^n dollars, say. A game is said to be
fair if the "entrance fee" is equal to the expectation of
the gain.

Clearly, N is the waiting time with geometric distribution, and we consider gain (cost function) of the form
$\phi(N) = k^N$ for fixed $k > 1$. The average gain is

$$\mathbb{E}\phi(N) = \sum_{n=1}^{\infty} k^n(1 - p)p^{n-1} = \sum_{n=1}^{\infty} (1 - p)/p = \infty.$$

Thus, a "paradox"—one would be willing to pay any sum, however large, for the privilege of participating in such a game. As a description of behavior, this is silly. Moreever, in those days infinite averages created uneasy feelings.

As a way out, Bernoulli proposed a "moral expectation," which according to him better described human behavior and is finite. In modern terminology, he introduced the utility function of the form $U = \log \phi(N) = N \log k$, whose mean is

$$\mathbb{E}\, U = (\log k)\, \mathbb{E}\, N = k \log k < \infty.$$

We can safely discard discussion of how moral this approach may be.

(b) <u>Rule of succession</u> (Laplace, 1812): Using the "principle of insufficient reason," Laplace developed his rule, which for long time was accepted as a basis for reliable scientific prediction. According to this rule, if an event of a certain kind has been observed n times in succession, then the probability of its occurrence at the next trial is $(n + 1)/(n + 2)$, approximately.

We skip with no regret the derivation, which is actually an exercise in the use of the "Bayes theorem" (i.e., conditioning), as we are more concerned with the non-mathematical implications resulting from the use of the rule in circumstances where it does not apply. The so-called Bayes hypothesis identified the lack of knowledge, or ignorance, with the notion of equiprobable events. We do not wish to open a Pandora's box of Bayes controversy and instead refer the reader to the well-known book by Feller for a clear discussion. We may remark here, however, that Laplace himself computed the probability (using his rule) that the sun will rise tomorrow, given that it has risen daily for n = 5000 years (since creation!). Feller added sarcastically that we are in the better position since regular service has followed for over one-and-half centuries.

(c) "Proof" of creation (Leibniz, 1713): In order to find the sum of the infinite series $1 - 1 + 1 - \cdots$, the famous philosopher and co-inventor of calculus G. W. Leibniz proposed the following solution. Observing that partial sums are alternatively 1 and 0, he stated that it is reasonable to assume that 1 and 0 are equiprobable. Hence, it suffices to take their average

$$1 \cdot (1/2) + 0 \cdot (1/2) = 1/2$$

as the "sum" of the series $1 - 1 + 1 - \cdots$.

In Leibniz's philosophy (employing a binary system), 1 represented the Supreme Being and 0 just plain Nothing. Then, the above argument might be interpreted as "proof" of creation: God created something (i.e., 1/2) from nothing.

The idea was so pleasing to Leibniz that he communicated it to the Jesuit Grimaldi, a missionary in China, in the hope that this "proof" would convert to Christianity the Emperor of China. Well, the Emperor did not believe in Operations Research.

It was Laplace who criticized Leibniz's argument in his "A philosophical essay of Probabilities" (reprinted by Dover). In view of our earlier remarks on the rule of succession, the following quotation from Laplace's essay may be appropriate: "I report this incident only to show to what extent the prejudices of infancy can mislead the greatest men."

In our modern times we may afford to add some more confusion to the story. Consider the geometric series $1 + \rho + \rho^2 + \cdots$, whose sum is of course $(1 - \rho)^{-1}$. Then, put $\rho = -1$ (why not?), so you will get again $1/2$ for that sum $1 - 1 + 1 - \cdots$. If you cannot figure your way out of this predicament, go back to square one! Or if you wish, you may look up the method of summation of divergent series, proposed by Cesaro in the 19th century, which again yields $1/2$ for our series.

****

We now close our discussion with the following quotation from the writings of Saint Augustine:

The good Christian should beware the mathematicians and all those who make empty prophesies. The danger already exists that mathematicians have made a convenant with the devil to darken the spirit and to confine man in the bonds of hell.

Chapter 5: Problems

1. Show again that the sample mean $\overline{X}$ of a random sample of size n with common mean μ (and known variance σ2) is the unbiased and consistent estimator of the population mean $\mu = \mathbb{E} X$ (see Subsection 10.2).

2. Show that the sample variance S^2, defined in Example 2 in Section 27, is the biased estimator of the pupulation variance $\sigma^2 = \text{var } X$, but the estimator $S^2 n/(n-1)$ is unbiased.

3. In Section 28 the MLE of λ in exponential distribution has been found to be $1/\overline{X}$. Show that

$$\mathbb{E}(1/\overline{X}) = \lambda n/(n - 1) \qquad \text{for } n > 1.$$

For n = 1, see Example 7 in Section 7.

4. Let $(X_1,\ldots,X_n)$ be a random sample from gamma distribution with parameters (λ, γ), where λ is unknown but γ is known (see Problem 61 in Chapter 1). Show that the method of moments estimator and the MLE of λ coincide and are equal to

$$\hat{\lambda} = \gamma/\overline{X},$$

where $\overline{X}$ is the sample mean. Show that

$$\mathbb{E}\,\hat{\lambda} = \lambda\gamma n/(\gamma n - 1) \qquad \text{for } \gamma n > 1.$$

What about estimation of γ?

5.   Let $(X_1, \ldots, X_n)$ be a random sample from distribution with density of the form

$$f(x) = ax^{a-1} \qquad \text{where } 0 < x < 1,\ a > 0,$$

and a is unknown. Show that

(i)   The MLE of a is given by $\hat{a} = -n/\log(X_1 \cdots X_n)$.

(ii)   The method of moments estimator is given by $\hat{a} = \overline{X}/(1 - \overline{X})$, where $\overline{X}$ is the sample mean.

6.   Provide details of calculations in Section 28 for

(i)   Gaussian example.

(ii)   Uniform example.

7.   Let $(X_1, \ldots, X_n)$ be a random sample of size n of r.v. X with mean μ and (known) variance σ^2. Consider an estimator of μ defined by

$$W = a_1 X_1 + \cdots + a_n X_n,$$

where $a_i \geq 0$ are constants such that $a_1 + \cdots + a_n = 1$. Show that:

(i)   W is unbiased.

(ii) $\operatorname{var} W \geq \operatorname{var} \overline{X}$, where $\overline{X}$ is the sample mean.

8.   Let $(X_1, \ldots, X_n)$ be a random sample of size n of r.v. X with mean μ and (known) variance σ^2. Define

$$Y_i = X_1 + \cdots + X_i \qquad \text{for } i = 1,\ldots,n$$

and consider an estimator of μ of the form

$$Z = \frac{Y_1 + \cdots + Y_n}{1 + 2 + \cdots + n}.$$

Show that Z is unbiased and

$$\text{var } Z = \frac{2}{3} \cdot \frac{2n + 1}{n + 1} \cdot \frac{\sigma^2}{n} > \sigma^2/n.$$

(<u>Hint</u>: use Problem 7.)

9. Let $(X_1,\ldots,X_n)$ be a random sample of size n with common variance σ^2.

    (i)   Show that

$$W = (1/n) \sum_{i=1}^{n} X_i(X_i - X_{n+1-i})$$

    and

$$Z = (2n)^{-1} \sum_{i=1}^{n} (X_i - X_{n+1-i})^2$$

are unbiased estimators of σ^2 when n is even.

    (ii)   What happens when n is odd?

10. Let $(X_1,\ldots,X_n)$ be a random sample from the distribution density

$$f(x) = (1/2)\, e^{-|x-\theta|}, \qquad -\infty < x < \infty,$$

where θ is an unknown parameter. Show that the MLE of θ is given by the median of the sample.

Note: Re-order the random sample according to magnitude ("order statistic")

$$X_{(1)} < X_{(2)} < \cdots X_{(n)}$$

where in particular $X_{(1)} = \min (X_1,\ldots,X_n)$ and $X_{(n)} = \max (X_1,\ldots,X_n)$. Then, the median is given by $X_{(m)}$, where $m = (n + 1)/2$ for n odd and by $(X_{(m)} + X_{(m+1)})/2$ where $m = n/2$ for n even.

11. Consider the simple queue, and suppose that n observations were made at instants of arrivals of n customers. Let $y_1,\ldots,y_n$ be observed numbers of customers in the system, and let $w_1,\ldots,w_n$ be observed waiting times of these incoming n customers. Assume that the system is in equilibrium. Using the method of moments show that estimates of intensities λ and μ are given by

$$\lambda = y^2(1 + y)^{-1}w^{-1}, \qquad \mu = y/w,$$

where y and w are values of respective sample means.

12. With reference to the sample function density g_t in Subsection 29.2, show that

$$\int \cdots \int \lambda^n e^{-\lambda t}\, ds_1 \cdots ds_n = (\lambda t)^n e^{-\lambda t}/n!,$$

where integration is taken over the region $0 \le s_1 \le \cdots s_n \le t$ (fixed t).

13. Suppose that the Poisson input is observed until a fixed number k of arrivals is recorded. Let $S = S(k)$ be the total time for these observations. Repeat the

procedure n times, and treat the result as a random sample $(S_1, \ldots, S_n)$ of size n. Show that the MLE of λ is

$$\hat{\lambda} = k/\bar{S}$$

where $\bar{S}$ is the sample mean.

14. Give the plausible heuristic argument that duration of a state i in a birth and death process should be exponential with parameter q(i). See Section 30.

15. Show that integrating the sample function density L_n stated in Subsection 30.2, over all $s_1 < \cdots < s_n < t$ (fixed t), would yield the joint distribution of N_t and $Y_0, \ldots, Y_n$.

16. Let g and h be two densities of some life times, with means m_1 and m_2, respectively. Let X be a life time with density f defined by:

$$f(x) = a\ g(x) + (1 - a)\ h(x), \qquad 0 \le x < \infty,$$

where a is a parameter such that $0 \le a \le 1$. Suppose that we wish to estimate a, assuming that g and h are known.

(i) Show that the method of moments yields the estimator of a given by:

$$\hat{a} = (\bar{X} - m_2)/(m_1 - m_2), \qquad \text{provided } m_1 \ne m_2,$$

where $\bar{X}$ is the sample mean. Verify that $\hat{a}$ is the unbiased estimator of a.

(ii) What about MLE? Write down the likelihood function $L(a)$, and comment on the difficulty of the problem.

(iii) Let g and h be exponential densities with parameters λ and μ, respectively. Assume that $\rho = \lambda/\mu < 1$, and take $n = 2$ (sample size 2). Write the explicit form of the likelihood function $L(a)$, and try to maximize it. Consider cases when the observed values x are smaller or larger than x_0 given by $\log \rho = -\mu(1 - \rho)x_0$.

Arithmetic progression:

$$a_n = a_1 + (n-1)h, \qquad a_1 + \ldots + a_n = \frac{n}{2}(a_1 + a_n)$$

$$1 + 2 + \ldots + n = \frac{1}{2}n(n+1)$$

$$1^2 + 2^2 + \ldots + n^2 = \frac{1}{6}n(n+1)(2n+1).$$

Geometric progression:

$$1 + \alpha + \ldots + \alpha^n = \frac{1-\alpha^{n+1}}{1-\alpha}$$

$$\sum_{i=0}^{n} i\alpha^i = \frac{\alpha}{1-\alpha}\left[\frac{1-\alpha^n}{1-\alpha} - n\alpha^n\right]$$

arithmetic mean: $\quad a = \dfrac{a_1 + \ldots + a_n}{n}$

geometric mean: $\quad g = \sqrt[n]{a_1 \ldots a_n}$

harmonic mean: $\quad \dfrac{n}{h} = \dfrac{1}{a_1} + \ldots + \dfrac{1}{a_n}.$

Series:

geometric:

$$\sum_{n=0}^{\infty} \alpha^n = \frac{1}{1-\alpha} , \qquad |\alpha| < 1.$$

exponential:

$$e^x = \sum_{n=0}^{\infty} \frac{x^n}{n!} , \qquad \text{all} \quad x,$$

$$\sin x = \sum_{n=1}^{\infty} (-1)^{n-1} \frac{x^{2n-1}}{(2n-1)!} , \qquad \cos x = \sum_{n=0}^{\infty} (-1)^n \frac{x^{2n}}{(2n)!} ,$$

$$\text{all} \quad x,$$

$$\sinh x = \sum_{n=1}^{\infty} \frac{x^{2n-1}}{(2n-1)!} , \qquad \cosh x = \sum_{n=0}^{\infty} \frac{x^{2n}}{(2n)!} ,$$

$$\text{all} \quad x,$$

$$e^{-x^2} = \sum_{n=0}^{\infty} (-1)^n \frac{x^{2n}}{n!} , \qquad \text{all} \quad x.$$

logarithmic:

$$\log(1+x) = \sum_{n=1}^{\infty} (-1)^{n-1} \frac{x^n}{n} , \qquad |x| < 1$$

$$\frac{1}{2} \log \frac{1+x}{1-x} = \sum_{n=1}^{\infty} \frac{x^{2n-1}}{2n-1} , \qquad |x| < 1.$$

binomial expansion:

$$(a+b)^n = \sum_{k=0}^{n} \binom{n}{k} a^k b^{n-k} , \qquad n = 0,1,\ldots .$$

binomial series:

$$(1+x)^{\alpha} = \sum_{n=0}^{\infty} \binom{\alpha}{n} x^n, \qquad \alpha \text{ real}, \quad |x| < 1,$$

$$(1+x)^{-\alpha} = \sum_{n=0}^{\infty} (-1)^n \binom{\alpha+n-1}{n} x^n, \qquad \alpha > 0, \quad |x| < 1,$$

$$(1-x)^{-1} = 1 + x + x^2 + x^3 + \ldots,$$

$$(1-x)^{-2} = 1 + 2x + 3x^2 + 4x^3 + \ldots, \quad |x| < 1.$$

polynomial expansion:

$$(a_1 + \ldots + a_k)^n = \sum \frac{n!}{r_1! \ldots r_k!} a_1^{r_1} \ldots a_k^{r_k}, \quad r_1 + \ldots + r_k = n.$$

harmonic series:

$$1 + \frac{1}{2} + \frac{1}{3} + \frac{1}{4} + \ldots \text{ diverges;} \qquad \sum_{n=1}^{\infty} \frac{1}{n^2} = \frac{\pi^2}{6}.$$

Binomial coefficients:

$$\binom{\alpha}{k} = \frac{\alpha(\alpha-1)\ldots(\alpha-k+1)}{k!}, \qquad \alpha \text{ real}, \quad k = 0,1,2,\ldots$$

for n positive integer or zero: $\binom{n}{k} = \frac{n!}{k!(n-k)!}$

$$\binom{n}{k} = \binom{n}{n-k}, \quad \binom{n}{0} = 1 = \binom{n}{n},$$

$$\binom{n}{k} = 0 \quad \text{for} \quad k > n \quad \text{and} \quad k < 0$$

$$\binom{n}{k-1} + \binom{n}{k} = \binom{n+1}{k}$$

$$\binom{-n}{k} = (-1)^k \binom{n+k-1}{k} \quad \text{for} \quad n > 0$$

$$\sum_{k=0}^{n} \binom{n}{k} = 2^n, \qquad \sum_{k=0}^{n} (-1)^k \binom{n}{k} = 0, \qquad \sum_{k=0}^{n} \binom{n}{k}^2 = \binom{2n}{n}$$

$$\sum_{k=1}^{n} (-1)^{k-1} \frac{1}{k} \binom{n}{k} = 1 + \frac{1}{2} + \ldots + \frac{1}{n}$$

Stirling formula:

$$n! \quad \sim \quad (2\pi)^{\frac{1}{2}} n^{n+\frac{1}{2}} e^{-n}$$

$$\lim_{n\to\infty} \binom{n}{k} x^n = 0, \qquad |x| < 1$$

$$\lim_{n\to\infty} \left(1 + \frac{x}{n}\right)^n = e^x, \qquad \lim_{x\to\infty} \frac{(1+x)^a - 1}{x} = a.$$

Special Functions:

Gamma functions:

$$\Gamma(z) \quad = \quad \int_0^\infty t^{z-1} e^{-t}\, dt \qquad (\text{Re } z > 0)$$

$\Gamma(z+1) = z\Gamma(z)$

For n positive integer: $\Gamma(n) = (n-1)!, \quad \Gamma(1) = 1,$

$\Gamma(2) = 1, \quad \Gamma(\frac{1}{2}) = \sqrt{\pi},$

$\Gamma(z+n) = (z+n-1) \ldots (z+1) z\Gamma(z)$

$$\frac{d^2}{dz^2} \log \Gamma(z) = \sum_{n=0}^{\infty} \frac{1}{(z+n)^2} \; ;$$

Euler constant $\gamma = -\dfrac{d}{dz} \log \Gamma(z+1) \Big|_{z=0}$

$$= \lim_{n\to\infty} \left(1 + \frac{1}{2} + \ldots + \frac{1}{n} - \log n \right)$$

$$= 0.5772157\ldots .$$

Incomplete gamma function:

$$\Gamma_y(z) = \int_0^Y t^{z-1} e^{-t} \, dt \qquad (\text{Re } z > 0)$$

$$\Gamma_y(z+1) = z\Gamma_y(z) - e^{-Y} y^z \qquad (z \neq 0),$$

$$\frac{1}{(n-1)!} \, \Gamma_y(n) = e^{-Y} \sum_{j=n}^{\infty} \frac{y^j}{j!} \qquad (n \text{ integer}).$$

Beta function:

$$B(p,q) = \int_0^1 x^{p-1}(1-x)^{q-1} \, dx, \qquad (\text{Re } p > 0, \quad \text{Re } q > 0)$$

$$B(p,q) = B(q,p), \qquad pB(p,q+1) = qB(p+1,q)$$

$$B(p,q) = \frac{\Gamma(p)\Gamma(q)}{\Gamma(p+q)} .$$

Incomplete beta function:

$$B_y(p,q) = \int_0^Y x^{p-1}(1-x)^{q-1} \, dx$$

$$(\text{Re } p > 0, \quad \text{Re } q > 0, \quad 0 \leq y \leq 1)$$

Binomial Distribution

$$p(j) \quad = \quad \binom{n}{j} p^j (1-p)^{n-j}$$

n	j	0.10	0.25	0.40	0.50	p
1	0	.9000	.7500	.6000	.5000	
	1	.1000	.2500	.4000	.5000	
2	0	.8100	.5625	.3600	.2500	
	1	.1800	.3750	.4800	.5000	
	2	.0100	.0625	.1600	.2500	
3	0	.7290	.4219	.2160	.1250	
	1	.2430	.4219	.4320	.3750	
	2	.0270	.1406	.2880	.3750	
	3	.0010	.0156	.0640	.1250	
4	0	.6561	.3164	.1296	.0625	
	1	.2919	.4219	.3456	.2500	
	2	.0485	.2109	.3456	.3750	
	3	.0035	.0469	.1536	.2500	
	4	.0001	.0039	.0256	.0625	
5	0	.5905	.2373	.0778	.0312	
	1	.3280	.3955	.2592	.1562	
	2	.0729	.2637	.3456	.3125	
	3	.0081	.0879	.2304	.3125	
	4	.0004	.0146	.0768	.1562	
	5	.0000	.0010	.0102	.0312	

Poisson Distribution

$$p(j) \ = \ \frac{\mu^j}{j!} \ e^{-\mu}, \qquad\qquad F(j) \ = \ \sum_{i=0}^{j} p(i)$$

	$\mu = 0.1$		$\mu = 0.2$		$\mu = 0.3$		$\mu = 0.4$		$\mu = 0.5$	
j	p(j)	F(j)	p(j)	F(j)	p(j)	F(j)	p(j)	F(j)	p(j)	F(j)
	0.		0.		0.		0.		0.	
0	9048	0.9048	8187	0.8187	7408	0.7408	6703	0.6703	6065	0.6065
1	0905	0.9953	1637	0.9825	2222	0.9631	2681	0.9384	3033	0.9098
2	0045	0.9998	0164	0.9989	0333	0.9964	0536	0.9921	0758	0.9856
3	0002	1.0000	0011	0.9999	0033	0.9997	0072	0.9992	0126	0.9982
4	0000	1.0000	0001	1.0000	0003	1.0000	0007	0.9999	0016	0.9998
5							0001	1.0000	0002	1.0000

	$\mu = 0.6$		$u = 0.7$		$u = 0.8$		$\mu = 0.9$		$\mu = 1$	
j	p(j)	F(j)	p(j)	F(j)	p(j)	F(j)	p(j)	F(j)	p(j)	F(j)
	0.		0.		0.		0.		0.	
0	5488	0.5488	4966	0.4966	4493	0.4493	4066	0.4066	3679	0.3679
1	3293	0.8781	3476	0.8442	3595	0.8088	3659	0.7725	3679	0.7358
2	0988	0.9769	1217	0.9659	1438	0.9526	1647	0.9371	1839	0.9197
3	0198	0.9966	0284	0.9942	0383	0.9909	0494	0.9865	0613	0.9810
4	0030	0.9996	0050	0.9992	0077	0.9986	0111	0.9977	0153	0.9963
5	0004	1.0000	0007	0.9999	0012	0.9998	0020	0.9997	0031	0.9994
6			0001	1.0000	0002	1.0000	0003	1.0000	0005	0.9999
7									0001	1.0000

μ = 1.5		μ = 2		μ = 3		μ = 4		μ = 5	
p(j)	F(j)	p(j)	F(j)	p(j)	F(j)	p(j)	F(j)	p(j)	F(j)
0.		0.		0.		0.		0.	
2231	0.2231	1353	0.1353	0498	0.0498	0183	0.0183	0067	0.0067
3347	0.5578	2707	0.4060	1494	0.1991	0733	0.0916	0337	0.0404
2510	0.8088	2707	0.6767	2240	0.4232	1465	0.2381	0842	0.1247
1255	0.9344	1804	0.8571	2240	0.6472	1954	0.4335	1404	0.2650
0471	0.9814	0902	0.9473	1680	0.8153	1954	0.6288	1755	0.4405
0141	0.9955	0361	0.9834	1008	0.9161	1563	0.7851	1755	0.6160
0035	0.9991	0120	0.9955	0504	0.9665	1042	0.8893	1462	0.7622
0008	0.9998	0034	0.9989	0216	0.9881	0595	0.9489	1044	0.8666
0001	1.0000	0009	0.9998	0081	0.9962	0298	0.9786	0653	0.9319
		0002	1.0000	0027	0.9989	0132	0.9919	0363	0.9682
				0008	0.9997	0053	0.9972	0181	0.9863
				0002	0.9999	0019	0.9991	0082	0.9945
				0001	1.0000	0006	0.9997	0034	0.9980
						0002	0.9999	0013	0.9993
						0001	1.0000	0005	0.9998
								0002	0.9999
								0000	1.0000

Exponential Function

x	e^x	e^{-x}	x	e^x	e^{-x}
0.00	1.0000	1.000000	1.00	2.7183	0.367879
0.05	1.0513	0.951229	1.05	2.8577	0.349938
0.10	1.1052	0.904837	1.10	3.0042	0.332871
0.15	1.1618	0.860708	1.15	3.1582	0.316637
0.20	1.2214	0.818731	1.20	3.3201	0.301194
0.25	1.2840	0.778801	1.25	3.4903	0.286505
0.30	1.3499	0.740818	1.30	3.6693	0.272532
0.35	1.4191	0.704688	1.35	3.8574	0.259240
0.40	1.4918	0.670320	1.40	4.0552	0.246597
0.45	1.5683	0.637628	1.45	4.2631	0.234570
0.50	1.6487	0.606531	1.50	4.4817	0.223130
0.55	1.7333	0.576950	1.55	4.7115	0.212248
0.60	1.8221	0.548812	1.60	4.9530	0.201897
0.65	1.9155	0.522046	1.65	5.2070	0.192050
0.70	2.0138	0.496585	1.70	5.4749	0.182631
0.75	2.1170	0.472367	1.75	5.7540	0.173774
0.80	2.2255	0.449329	1.80	6.0498	0.165299
0.85	2.3396	0.427415	1.85	6.3598	0.157237
0.90	2.4596	0.406570	1.90	6.6859	0.149569
0.95	2.5857	0.386741	1.95	.0237	0.142274
			2.00	7.3891	0.135335

The Normal Distribution

$$\phi(t) = \frac{1}{\sqrt{2\pi}} e^{-\frac{t^2}{2}}, \qquad \Phi(t) = \int_{-\infty}^{t} \phi(s)\, ds$$

t	$\phi(t)$	$\Phi(t)$	t	$\phi(t)$	$\Phi(t)$
0.0	0.398942	0.500000	2.3	0.028327	0.989276
0.1	.396952	.539828	2.4	.022395	.991802
0.2	.391043	.579260	2.5	.017528	.993790
0.3	.381388	.617911	2.6	.013583	.995339
0.4	.368270	.655422	2.7	.010421	.996533
0.5	.352065	.691462	2.8	.007915	.997445
0.6	.333225	.725747	2.9	.005953	.998134
0.7	.312254	.758036	3.0	.004432	.998650
0.8	.289692	.788145	3.1	.003267	.999032
0.9	.266085	.815940	3.2	.002384	.999313
1.0	.241971	.841345	3.3	.001723	.999517
1.1	.217852	.864334	3.4	.001232	.999663
1.2	.194186	.884930	3.5	.000873	.999767
1.3	.171369	.903200	3.6	.000612	.999841
1.4	.149727	.919243	3.7	.000425	.999892
1.5	.129518	.933193	3.8	.000292	.999928
1.6	.110921	.945201	3.9	.000199	.999952
1.7	.094049	.955435	4.0	.000134	.999968
1.8	.078950	.964070	4.1	.000089	.999979
1.9	.065616	.971283	4.2	.000059	.999987
2.0	.053991	.977250	4.3	.000039	.999991
2.1	.043984	.982136	4.4	.000025	.999995
2.2	.035475	.986097	4.5	.000016	.999997

Appendix C

Suggestions for Further Reading

There are really no elementary texts on stochastic processes. On the other hand, there is a great availability of books on statistics, and most of them contain chapters on probability and some of them contain a chapter or two on elementary stochastic processes (especially renewal and Markov chains).

Medium level books on stochastic processes:

1. N.T.J. Bailey, The Elements of Stochastic Processes with Applications to the Natural Sciences, J. Wiley, 1964.

2. E. Parzen, Stochastic Processes, Holden Day Inc., 1962.

3. S.M. Ross, Introduction to Probability Models, Academic Press, 1972.

4. See also articles in "Scientific American."

Advanced texts on stochastic processes:

5. A.T. Bharucha-Reid, Elements of the Theory of Markov Processes and their Applications, McGraw-Hill, 1960.

6. E. Cinlar, Introduction to Stochastic Processes, Prentice-Hall, Inc., 1975.

7. W. Feller, Introduction to Probability Theory and its Applications, Vol. 1, J. Wiley, 1957.

8.   S. Karlin, H.M. Taylor, A First Course in Stochastic
     Processes, Academic Press, 1975.

9.   N.U. Prabhu, Stochastic Processes, McMillan, 1965.

10. S.M. Ross, Applied Probability Models with Optimization
     Applications, Holden-Day, 1970.

There are also specialized works devoted to special
topics, like queueing, reliability, renewals as well as books
for mathematicians on Markov Processes, random walks, etc.
For a particularly successful blend of "ultra-advanced" theory
with significant practical applications, consult Volume 2 of
Feller.

The following are some recently published books.

11. S. Karlin, H. M. Taylor, A Second Course in Stochastic
     Processes, Academic Press, 1981. (This is the continua-
     tion of Ref. 8.)

12. H. M. Taylor, S. Karlin, An Introduction to Stochastic
     Modeling, Academic Press, 1984. (This is a baby ver-
     sion of Refs. 8 and 11.)

13. M. Hofri, Probabilistic Analysis of Algorithms,
     Springer-Verlag, 1987. (This is a rather specialized
     text, but it is highly recommended for its lucid style
     and concrete probabilistic applications.)

14. R. Syski, Introduction to Congestion Theory in Tele-
     phone Systems, North-Holland, 1986, second edition.
     (This revised version contains the additional chapter
     on Markovian Queues.)

15. J. Gani (ed.), The Craft of Probabilistic Modelling
     (A Collection of Personal Accounts), Springer-Verlag,
     1986.

16. T. L. Saaty, J. M. Alexander, Thinking with Models, Pergamon Press, 1981.

17. S. Kotz, N. L. Johnson, C. B. Read (eds.), Encyclopedia of Statistical Sciences, J. Wiley. Among numerous articles of general interest these two are humbly recommended:

    R. Syski, Multi-server Queues, Vol. 5, 1985,

    R. Syski, Stochastic Processes, Vol. 8, 1988,

and the excellent view of statistics:

    E. L. Lehmann, Statistics: An Overview, Vol. 8, 1988.

Index

Absorbing state, 289, 295
Accidents, 158
Age, see Life time
Approximation, 10, 85
 Gaussian, 83, 90
 Poisson, 153
Arrivals, see Input
Average, see Expectation
Average cost, see Cost

Backward equation, 293, 298
 308, 330, 333
Balance equation, 232, 276,
 285, 294, 297, 305, 380
Bernoulli trials, 133-141
Beta distribution, 127
Beta function, 397
Binomial coefficients, 135,
 136, 395
Binomial distribution, 136,
 166, 167, 292
 negative, 177
 truncated, 145
Binomial expansion 135, 394
Birth equations, 277
Birth process, 275, 289, 373
 life time in, 280-283
Birth and death equations,
 288, 298
Birth and death process,
 286, 361
 estimation in, 363-371
 intensities of, see Rate
 life time in, 294, 297
Birth and death solutions,
 291

Branching
 process, 303
 property, 309
 theorem, 314
Bus problem, 29, 32, 203,
 204, 206
Busy period, 249, 259, 294,
 302, 320

Central limit theorem, 84, 90
Chapman-Kolmogorov equation,
 232, 380
Chebyshev inequality, 87-89
Chi-square distribution, 117
Combination of life times,
 34, 42
Conditioning, 22-24, 159,
 238, 240, 285, 365, 380,
 381
Convolution, 40, 62, 176,
 198, 310
Cost
 average, 52, 58, 76
 declining, 61
 function, 52, 58
 linear, 53, 58
 quadratic, 54
Creation, 385
Customers, 233, 257, 321
Cut-off point, 55

Data, 344

Death process, 303, 331
Deconvolution, 129
Density, 12, 18, 75
 conditional, 27
 convolution of, 40
 joint, 34
 marginal, 35
 renewal, 196
 sample function, 358, 365
 symmetric, 107
Difference-differential
 equations, 236, 268, 277,
 289
Differential equations, 243,
 269, 311
Discount, 61
Distributions
 continuous, 16, 75
 discrete, 131, 263, 290
Distribution function
 (d.f.), 14, 18, 75, 138,
 155
 complementary, 15, 18
 conditional, 26
 joint, 35
 marginal, 36

Equilibrium, 204, 247, 263,
 379
 equations, 264, 290
Erlang distribution, 157,
 174
ESP, 144
Estimate, 346
Estimation, 343, 349, 356
Estimator, 345
 consistent, 347
 maximum likelihood, 352
 method of moments, 348
 unbiased, 346
Event, 3,6
 complementary, 6
 sure, 5
 total, 5
Events
 combination of, 6, 15, 22,
 35, 37, 211
 determined by random variables, 11, 12, 40
 independent, 37
Expectation, 19, 21

(Expectation)
 of compound random variable,
 160-164, 178, 240, 318
 conditional, 238, 312, 333,
 381
 infinite, 56, 214, 282, 320
 of products, 59
 of sum, 41, 44
Explosion, 282, 314
Exponential distribution, 17,
 21, 25, 48, 193, 280, 361,
 391
Exponential series, 154, 394
Extinction, 294, 296, 313,
 314
Extreme life times, 45-47,
 212, 316

Factorial, 135
Failure, 132
First crossing, 207, 212
 entrance, 256, 297, 312, 333
 passage, 251, 256
 return, 256
Forward equations, 289, 294,
 308, 333
Frequency, 8, 141
Friends meeting, 68

Gamma distribution, 126
Gamma function, 396
Gaussian distribution, see
 Normal distribution
Generating function (cost
 function), 170, 175-177,
 268, 310
Geometric distribution, 147
 169-171, 208, 265, 319
Geometric series, 148, 394
Group input, 321
Group of lines, 253-256, 327

Hazard function, 28
Holding time, 240

Idle period, 249, 259

Independence, 36, 39, 67
Iid, 88, 345
Inference, 344
Infinite number of lines,
  331
Initial distribution, 247,
  366
Input, 233, 260, 274
 finite, 330
 group, 321
 in random time, 240
 Poisson, see Poisson
  process
 process, see Birth process,
  Renewal process
Integral, 13-14
 double, 66, 67, 73
Integral equations, 197,
  212, 220
Intensity, see Rate
Interarrival time, 29, 237,
  258, 359, 363

Joint distribution, destiny,
  see Distribution function,
  destiny
Jumping rabbit, 206

Killing, 297
Kinetic energy, 54

Laplace transforms, 61, 197,
  278
 solutions by, 198, 199,
  213, 278
Law of large numbers, 8, 89,
  140
Learning model, 242, 256
Life time, 9, 19
 average, see Expectation
 constant, 33
 continuous, 16
 discrete, 131
 exponential, see Exponential distribution
 extreme, 45
 finite, 11, 281

(Life time)
 independent, see Independence
 infinite, 56, 214, 282
 remaining (residual), 31,
  204
 total, 43, 88, 163, 176,
  204, 280, 363
 uniform, see Uniform distribution
Likelihood function, 350, 367
Limiting distribution
 for state, 256, 263, 279,
  290
 for time, 212, 220
Linear growth process, 294
Linear transformation, 62
Little formula, 328

Marginal distribution, see
  Distribution function
Markov process (property), 2
  231, 285, 361, 380
Martingale, 380
Maximum likelihood, see MLE
Mean, see Expectation
Median, 390
Memoryless property, 26, 148,
  231, 379
Method of moments, 348
Mixing, 163, 178
MLE, 351
Models, 1, 8, 284, 291, 376-
  377, 381
Moments, 20, 76, 119
 absolute, 80

N.E.D., see Exponential distribution
Normal distribution, 76-80
 folded, 116
 standard, 77, 114
Number of arrivals, 233, 238,
  240
 of renewals, 191
 in the system, 258, 285

Order statistic, 390
Overselling of tickets, 142

Parallel arrangement, 50
Parameters, 342
Pareto distribution, 98
Poisson distribution, 153,
   173-175, 195, 215-216,
   237, 292
 transform of, 279, 324
 truncated, 157
Poisson process (input),
   233, 237, 277, 289, 356,
   375
 estimation of, 357-360, 390
Poisson renewals, 195, 200
Points, distribution of, 59
Probability, 2-8, 40
 conditional, 22, 24
 theory, 8, 10
Probability of
 blocking, 146, 157, 339
 extinction, 295, 314
 first entrance, 207, 212,
   256, 299, 313
 prolongation, 24
 termination, 237, 261
Profit, 166
Prolongation of life time,
   22, 26
Proportion, 74

Quadratic equation, 70
Queue, 9, 256
 discipline, 257
 length, 258
 simple, 257-273, 372
 waiting, see Waiting time
Queueing intensity, 265

Random
 arrivals, see Input
 interval, 206
 process (fluctuations), see
   Stochastic process
 sample, 345
 sums, 162, 163, 178, 204

(Random)
 time, 239, 297
 variables, (r.v.), 10, 75,
   378, see also Life times
   compound, 162-164, 178,
   240, 309, 318
   independent, 36, 39, 67
 walk, 209, 213
Rate
 arrival, 234
 birth, 275
 birth and death, 287, 361
 failure, 28
 hazard, 27
 hindrance, 242
 learning, 242
 Poisson, 234
 renewal, 199
Recurrence relations (for
   distribution functions),
   190, 212
Reliability, 51
Renewal (s), 188
 age, 206
 density, 196-198
 equation, 196, 249
 function, 195-198
 Poisson, 195, 200
 process, 190, 191
  alternating, 249
 rate, 199
 theorem, 199, 204
 total time, 204
Rule of succession, 384

Sample
 function density, 358, 367
 mean, 89, 140, 247
 moments, 248
 random, 345
 variance, 247
Sampling with replacement,
   141
Series arrangement, 49
Service time, 258
Servicing of machines, 292
Server, 256
Slack period, 259
Standard deviation, 21
State, 234, 284
Stationary process, 379
Statistic, 345, 390

Statistical equilibrium, see
   Equilibrium
Statistics, 8, 342
Steady State, see Equilib-
   rium
Stirling formula, 152, 396
Stochastic process, 1, 231,
   378
Stopping time, see Random
   time
St. Petersburg paradox, 383
Strict order service, 257,
   266
Success, 132
Sums of random variables,
   88, 162, 176, 270, 363
Survival function, 15
System, 233

Taboo probability, 297, 313
Taboo set, 210, 297
Telephone models, 144, 291
Time
 continuous, 12, 131
 discrete, 12, 131
 first entrance, 297
 random, 239, 297
 spent in a state, 280, 361,
  391
 to extinction (absorption),
  303

total, 43
waiting, see Waiting time
Traffic, 145
Transition probability, 234,
   285, 380
 equations for, 277, 288,
   293, 308
Truncated binomial, Poisson,
   see Binomial, Poisson
   distributions
Transforms, see Laplace
   transform, Generating
   function
uniform distribution, 16,
   21, 355
utility, 52, 112

Variance, 20

Waiting
 line, see Queue
 room (finite), 338
 system, 9
 time, 30, 258, 266, 270,
  302, 332
  for bus, 30, 203, 204
  for completion, 45-46
  distribution, 267, 269,
   271, 272, 328, 332
  geometric, 146